Advanced Flow Cytometry:
Applications in Biological Research

Edited by

R. C. Sobti
Department of Bio-Technology,
Panjab University,
Chandigarh 160014
India

and

Awtar Krishan
Department of Radiation Oncology,
University of Miami Medical School,
Miami, FL 33101
U.S.A.

Reprinted from Methods in Cell Science, Volume 24(1–3), 2002

KLUWER ACADEMIC PUBLISHERS
DORDRECHT / BOSTON / LONDON

Table of Contents

A C.I.P. Catalogue record for this book is available from the Library of Congress.

ISBN 1-4020-1484-8

Published by Kluwer Academic Publishers,
P.O. Box 17, 3300 AA Dordrecht, The Netherlands

Sold and distributed in North, Central and South America
by Kluwer Academic Publishers,
101 Philip Drive, Norwell, MA 02061, U.S.A.

In all other countries, sold and distributed
by Kluwer Academic Publishers,
P.O. Box 322, 3300 AH Dordrecht, The Netherlands

Printed on acid-free paper

Printed in the Netherlands

Methods in Cell Science 24: 1–9 (2002).

Flow cytometry, an overview

L. Scott Cram
National Flow Cytometry Resource, Los Alamos National Laboratory, Los Alamos, New Mexico, 87545, USA

1. Introduction

Flow cytometry is a technique for making quantitative measurements on single cells or cellular constituents at very high rates of speed. The value of flow cytometry has been likened to finding a needle in a haystack. A measurement of the average property of a haystack would fail to identify the needle. One alternative is to measure each particle in the haystack one at a time using a parameter that would distinguish a piece of straw from something metallic. By measuring individual particles one can obtain a precise description of each particle from among a much larger population as illustrated in Figure 1.

In this analogy, the measurement of particle length would not be a very good descriptor, however measuring the magnetic property of each particle would provide excellent discrimination. Thus it is with flow cytometry; measurements are made on each individual particle within the suspension as contrasted with measuring an average property for the entire population. Average measurements fail to detect subpopulations that have small differences from the overall population. Particle fluorescence and light scatter are two of the most common measurements made in a flow cytometer. With those two descriptors in hand one can also sort individual particles (e.g., cells and chromosomes) to recover subpopulations or sub fractions from the larger pool of particles provided the measurement of individual particles adequately discriminates the unique particles from the remainder of the population.

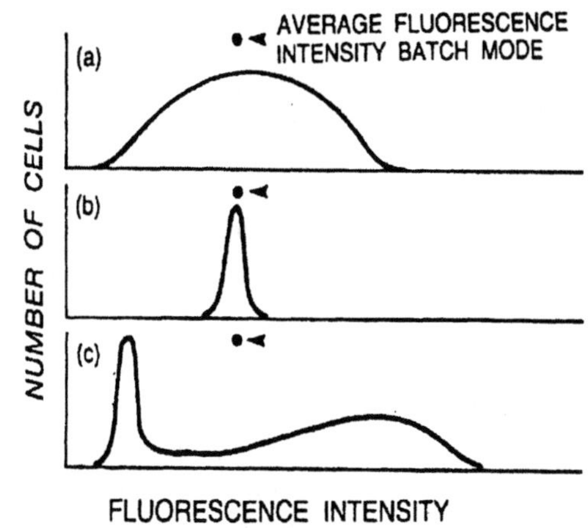

Figure 1. Flow cytometers can readily detect rare populations and can qualify differences between subpopulations without prior separation.

The number of applications of flow cytometry and particle sorting are numerous. Based on an ever-increasing number of fluorochromes with excellent specificity for cellular constituents, the field has expanded rapidly. Nucleic acid content, enzyme activity, calcium flux, membrane potential and intracellular pH are routine measurements. Conjugation of fluorescent dyes to polyclonal and monoclonal antibodies have made it possible to study whole blood cell subpopulations in great detail. By combining light scatter measurements with fluorescence one can further resolve the constituent parts of complex cell populations.

While flow cytometers have evolved into ever simpler machines, their inherent complexity should not be overlooked both in terms of their potential to make very sophisticated multivariate measurements and their requirement for excellent sample quality. Likewise the combinatorial selection of fluorochromes is also problematic. The quality of a sample will determine the ability to resolve subpopulations as well as the purity of a sorted fraction. The aim of sample preparation is to prepare a suspension of single particles that have been stained in a specific way to identify a particular characteristic; DNA or RNA content, specific proteins, etc. Single particles and minimal debris is necessary to ensure a sample flow stream that is uniform and without disrupting particles that create turbulence. Combinations of fluorochromes, while a powerful technique, must be considered carefully to avoid spectral emission overlap and energy transfer between closely bound fluorochromes. These problems can be successfully managed but must be considered when selecting fluorochromes for labeling cells or cellular constituents such as chromosomes.

The size ranges of particles that have been analyzed using the techniques of flow cytometry are enormous (Table 1). From zooplankton that are in the size range of a centimeter to single fluorescent molecules, the dynamic size range is over six orders of magnitude.

This overview of flow cytometry will cover instrumentation (lasers, fluidics, optics, electronics, and sorting), data display and analysis (single parameter histograms, multiparameter data, and list mode data and gating), and fluorescence measurements

Table 1. The size range of particles analyzed in a flow cytometer covers six orders of magnitude

Particle Size (microns)	'Particle'
100,000.	
	Larger plankton
10,000. = 1 cm	
	Zooplankton
1,000.	Algae
	Encapsulated cells/bacteria
100.	Multicellular spheroids, Bioreactors
	Cell aggregates, Plant protoplasts, Microenvironments
10.	Mammalian cells
	Microspheres, Cell nuclei
1.	Chromosomes, Bacteria
	Cellular vesicles, Picoplankton
0.1	Liposomes
	DNA fragments
0.01	Virus particles
0.001	Phycoerythrin
	Other individual molecules
0.0001 = 1 Å	

(excitation and emission, fluorochromes, filters, and color compensation).

2. Instrumentation

2.1. *Basic concepts*

The basic concept of a flow cytometer is illustrated in Figure 2. As a single particle intersects the central region of a laser beam(s) focused to provide uniform illumination across the diameter of the sample stream, fluorescence and light scatter properties of the particle are measured. Measurements consist of light scatter (particle size and surface characteristics), fluorescence (total intensity, maximum intensity, polarization, and lifetime), particle volume ('Coulter volume'), and specialty measurements such as phase shift (internal cell environment). With appropriate labeling reagents, these measurements provide information about the amount of DNA, RNA, protein, cell surface molecules, and other intracellular characteristics. The principles of fluid flow are used in the flow chamber to position cells in the central most region of the laser beam thereby assuring uniform illumination. By maintaining a flow stream diameter on the same size scale as the object being measured, it is possible to position cells like beads on a string when they arrive in the constricted region of the flow cell. Critical but seldom appreciated characteristics of this method of analysis include the nano-liter to pico-liter probe volume (the volume defined by the intersection of the sample stream and laser beam), and the measurement of only particle associated fluorescence. The measurement of only particle associated fluorescence eliminates the need to remove free fluorescence in solution. Single molecule fluorescence sensitivity has been achieved in some specialized systems. Fluorescence resolution of 1% can be achieved. Different sized particles containing different combinations of fluorochromes can be efficiently analyzed at rates of up to 10,000 cells per second.

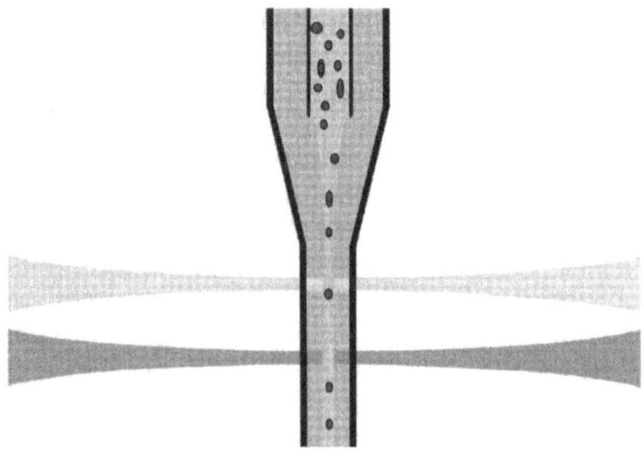

Figure 2. Particles flow through one or more laser beams where they are analyzed for fluorescence at multiple wavelengths and light scatter angles.

The fluorescence of each cell is collected using conventional optical components, broken into selected wavelength regions and focused onto an optical detector. From the optical detector (e.g., a photomultiplier) the fluorescence signal is converted to an electronic signal that is amplified, analyzed and stored in a dedicated computer. Figure 3 illustrates one method by which the optical signal is processed,

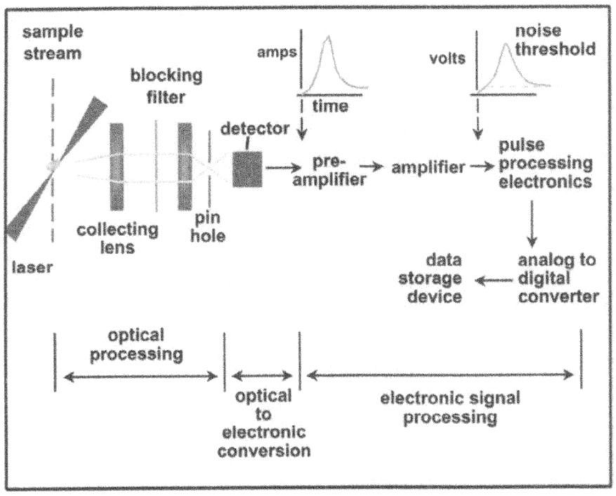

Figure 3. The optical and electronic processing of a fluorescence signal in a typical flow cytometer.

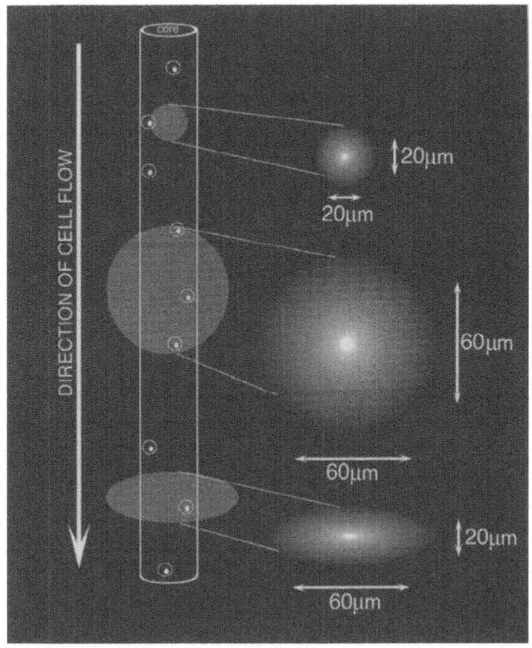

Figure 4. Two different options for beam shape illustrate the advantages of an elliptically shaped beam.

converted to an electronic signal and the signal processed.

2.2. *Laser light sources*

Most flow cytometers use one or more lasers as light sources to excite whatever fluorochrome(s) are bound to the cells. Lasers have the advantage of providing intense, single wavelengths for excitation and in turn simplify filter selection for fluorescence detection. Three of the most common lasers in commercial instruments and their most useful wavelengths are Argon ion (365 nm, 488 nm or 514 nm), Krypton (567 nm, 646 nm) and Helium-Neon (633 nm). Lasers also provide a polarized light beam that has proven useful for special applications where molecular orientation matters. Laser intensity is highest at the center of a focused laser beam. To provide uniform illumination across the sample stream, focusing optics are used to provide an elliptically focused beam, thus providing more uniform illumination across the diameter of the laser beam. Figure 4 illustrates three different beam shapes.

A spherical focus provides high beam intensity and fluorescence sensitivity but the beam intensity varies greatly across the sample stream diameter whereas a tight vertical focus provides adequate intensity and fluorescence sensitivity while achieving a few percent difference in intensity across the diameter of the sample stream.

2.3. *Fluidics*

The goal of the fluidics system is to achieve a steady flow rate and to regulate the sample to sheath flow ratio. The sample stream diameter is determined by the relative flow rate of the sheath stream that surrounds the sample stream and the sample stream flow rate. Major variables are sheath pressure, sample pressure and the size of the flow cell orifice. For most applications the flow cell orifice is around 75 μm. The issue of sample flow rate is critical and is illustrated in Figure 5.

At a low sample to sheath input rate (left figure), tight hydrodynamic focusing is achieved along with a lower coefficient of variation (CV) (tighter peaks) and lower coincidence rates. At high sample to sheath rates (right figure), poor hydrodynamic focusing results along with higher CV's (broader peaks) resulting in a high coincidence rate. Sample stream diameters should be set to the same diameter as the size of the particles being measured. For high-resolution chromosome analysis a sample stream diameter of less than 2 μm is used. A higher cell flow rate (cells per second) will increase the sample stream (core diameter) and decrease sample resolution. Sample concentration (cells/ml) therefore has a direct affect on sample resolution if a constant flow rate and stream rate are maintained as illustrated in Figure 6.

2.4. *Fluorescence and light scatter analysis*

Figure 7 illustrates the fluorescence and light scatter signals that are 'seen' by the respective detectors. For a cell illuminated by a blue laser light (488 nm) and stained with two dyes emitting red and green fluorescence), the optical channel sees red and green fluorescence and scattered blue laser light. The light scatter detector is masked to block the incident laser beam while still allowing a measurement of low angle scattered light.

Spectroscopic analysis of the multiple color

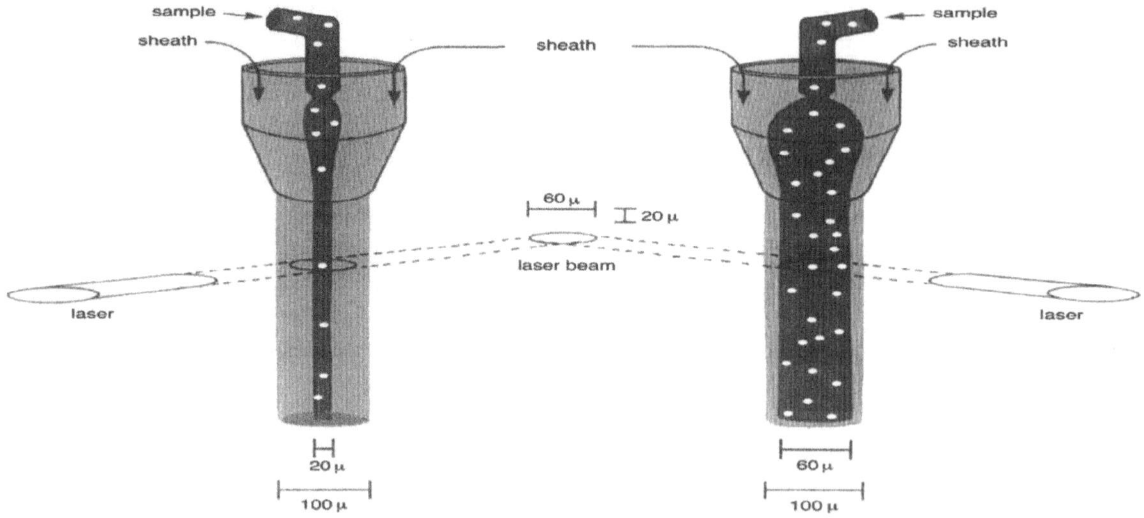

Figure 5. Relative sample to sheath rates determines the sample stream diameter.

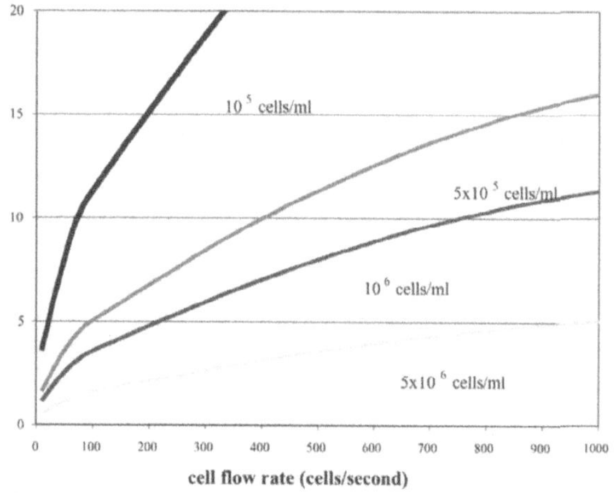

Figure 6. Core diameter (in microns) as a function of cell flow rate and cell concentration.

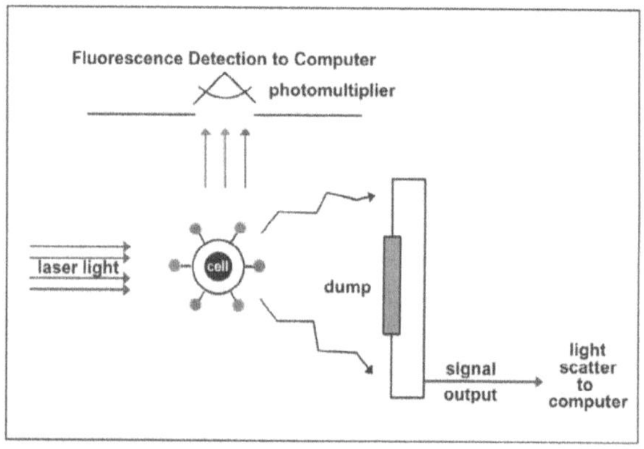

Figure 7. Fluorescence and light scatter detection.

emission is achieved using optical filters designed to block scattered laser light that enters the optical channel, which is positioned at right angles to the intersection of the laser beam and sample stream. This arrangement is illustrated in Figure 8. A combination of interference and absorbance filters is used to provide the required spectral resolution. A typical three-color analysis arrangement is illustrated where a short pass, long pass, band pass, and dichroic filter are used. In addition two light scatter signals are detected and recorded, one a forward scatter signal (FSC signal) and second, a side scatter signal (SSC signal).

2.5. *Signal amplification and dynamic range*

Photodetection has historically been achieved using high sensitivity, high amplification. Photomultiplier tubes continue to serve this purpose for most applications. New, solid state detectors are beginning to appear in some specialized flow cytometers. After conversion of the optical signal to an electronic signal, two types of amplifiers are available, linear and logarithmic. The data obtained using these two types of amplifiers is exactly the same, however, the distributions look different due to how the analog signal (the electronic pulse shape) is converted into a digital signal for data storage and display. This difference is illustrated in Figure 9. More important is the dynamic range that is achieved with a logarithmic amplifier. The choice of linear or logarithmic amplification depends on the type of biology one is analyzing. For example (Figure 10) if a limited dynamic range is needed, as is often the case with DNA analysis (a two fold difference in DNA content) then linear amplification is used.

However, surface marker expression covers a much broader dynamic range (100–10,000 fold) of fluorescence intensities, due in part to the large variation in cell surface area of cells in different phases of their life cycle and cell heterogeneity

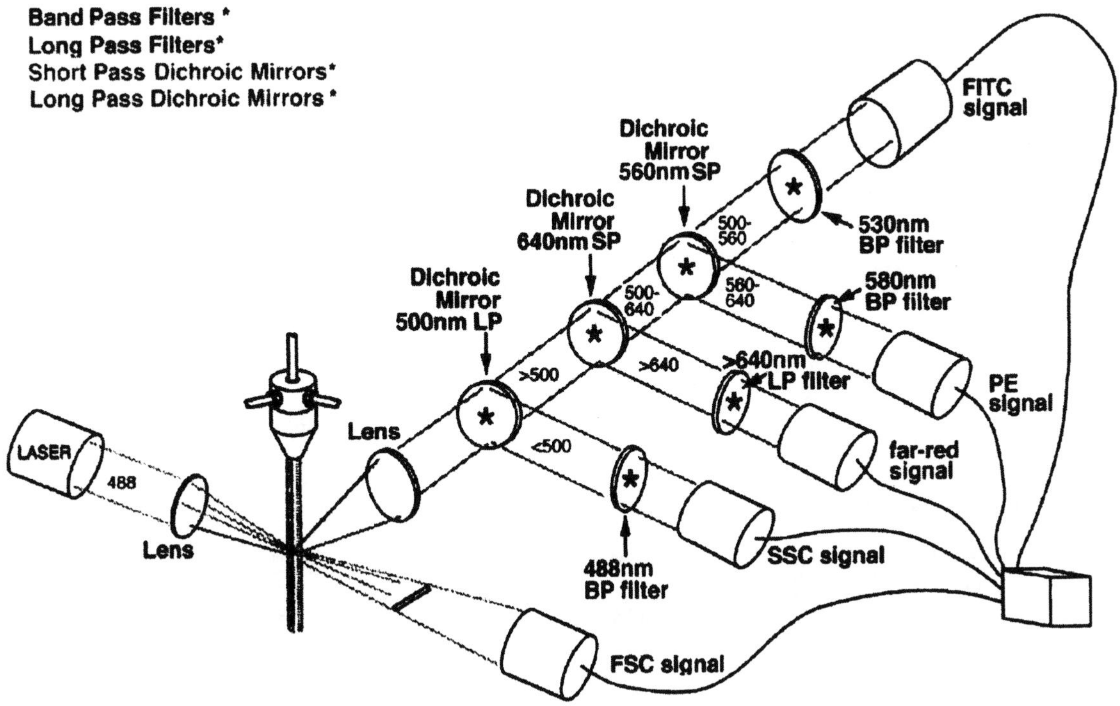

Figure 8. A typical single laser 3-color fluorescence and 2-angle light scatter set up.

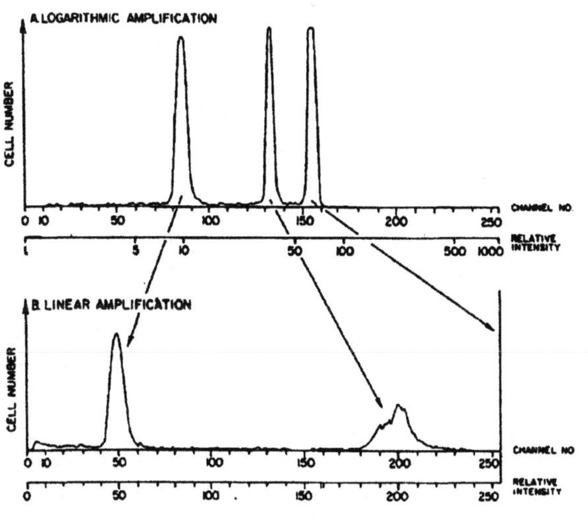

Figure 9. Comparison of logarithmic amplification with linear amplification.

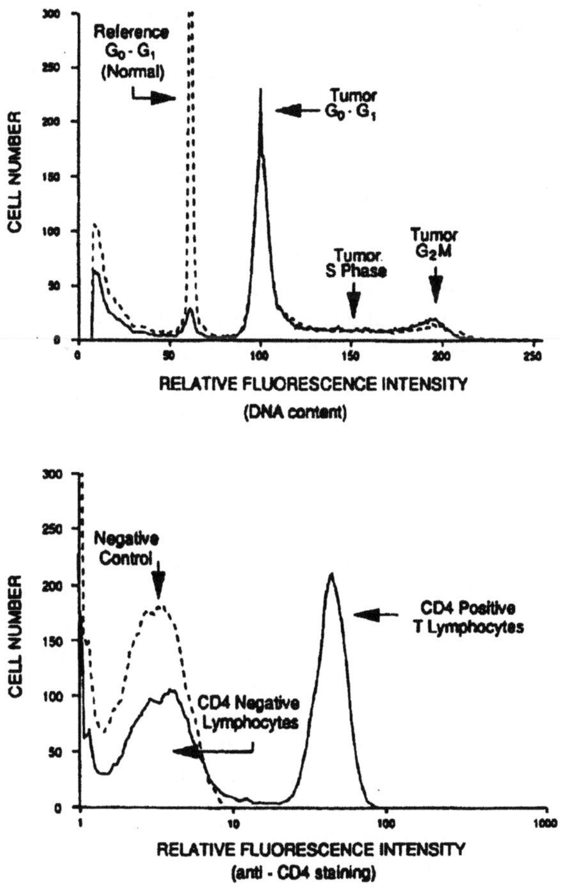

Figure 10. Cells analyzed for DNA content have a limited dynamic range of fluorescence intensity as contrasted with cell surface antigen content. Immunofluorescence has a very broad dynamic range of fluorescence intensity.

based on function and origin. Immunofluorescence is often measured using a logarithmic amplifier to capture the entire range of fluorescence intensities as demonstrated in Figure 10. Most logarithmic amplifiers are designed to cover four decades of dynamic range.

2.6. *Data display and analysis*

Single parameter data is readily analyzed using curve fitting routines available from any of several commercial sources. The analysis of data such as illustrated in Figure 11 has become increasingly

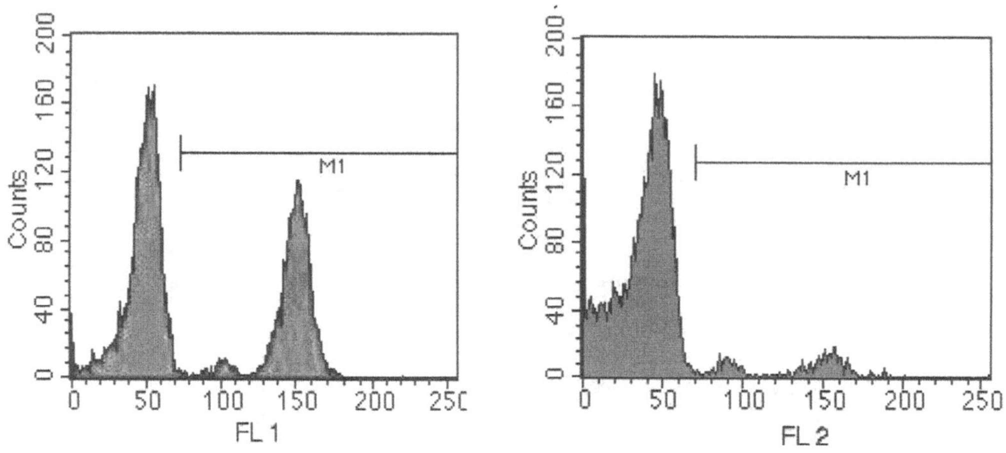

Total Events: 6000

Marker	Mean	Median	Peak Ch	% Positive
All	87.76	58.00	56	
M1	146.08	149.00	150	48%

Total Events: 6000

Marker	Mean	Median	Peak Ch	%Positive
All	41.32	42.00	0	
M1	134.06	148.00	156	12%

Figure 11. Single parameter data: typical immunofluorescence display and analysis.

sophisticated and very powerful. The data shown is a single parameter histogram of the number of events as a function of fluorescence (FL) intensity. Using a variety of parametric tools one can precisely determine peak position (mean and median), peak area, percentage of cells in different peaks, and the contribution of background events to the peak area. Data analysis consists of fitting Gaussian(s) and second order polynomials to derive the numerical values.

In this example, peripheral blood cells were stained with two different cell surface specific reagents, each labeled with a different fluorochrome. The two distributions were analyzed and displayed separately. Since the binding specificity identified two different subpopulations, the percentage positive cells in the two distributions was much different. In one case 48% of the cells were positive and in the other only 12% were positive based on the region (M1) defined as positive.

The presence of background counts can be an indication of cellular debris, subpopulations of cells, or incomplete staining. The shape of the background distribution can offer clues as to the origin of such events. Two-parameter data can be used to more completely distinguish among subpopulations. Correlated (that is critical) multiparameter data can further resolve populations assuming that the probes distinguish unique properties among different subpopulations. Figure 12 illustrates four different types of two-parameter data display.

Data analysis of two-parameter data often consists of integrating the total number of events within a defined area. Peak position can be determined by performing a one-dimensional collapse of the data onto a single axis followed by parametric analysis as described for Figure 11. Another option is to draw a contour line at the base of a distribution and integrate the total number of events within that contour. In some instances the position of the contour line that defines the demarcation between true counts and background events can be established as a percentage of the peak value. A cautionary note here is to be aware that some published articles artificially exclude background events. When data is displayed using an artificial background cut-off, the true overlap between populations cannot be observed.

Data analysis of background counts in two-parameter data displays (Figure 12) is problematic when the shape of the distribution is irregular. Contour plots have proven to be the most useful method of two-parameter data display. The centroid of two-parameter data is sometimes used to determine relative fluorescent intensity of each population. A '3D' plot can be useful for visualizing peak resolution. Such visualization is achieved by rotating the distribution to view peaks from different angles and is therefore seldom seen in printed publications. Depending on the diversity of the population being measured, contour plots can be any combination of log and linear displays.

2.7. *Light scatter properties*

Light scattering properties reflect intrinsic physical differences among cells. Light scatter (see Figure 7) is generally measured in the forward (low angle) direction from the laser beam direction and is strongly influenced by cell size. Right angle scatter (measured at 90° from the direction of the laser beam) has been shown to reflect granularity or

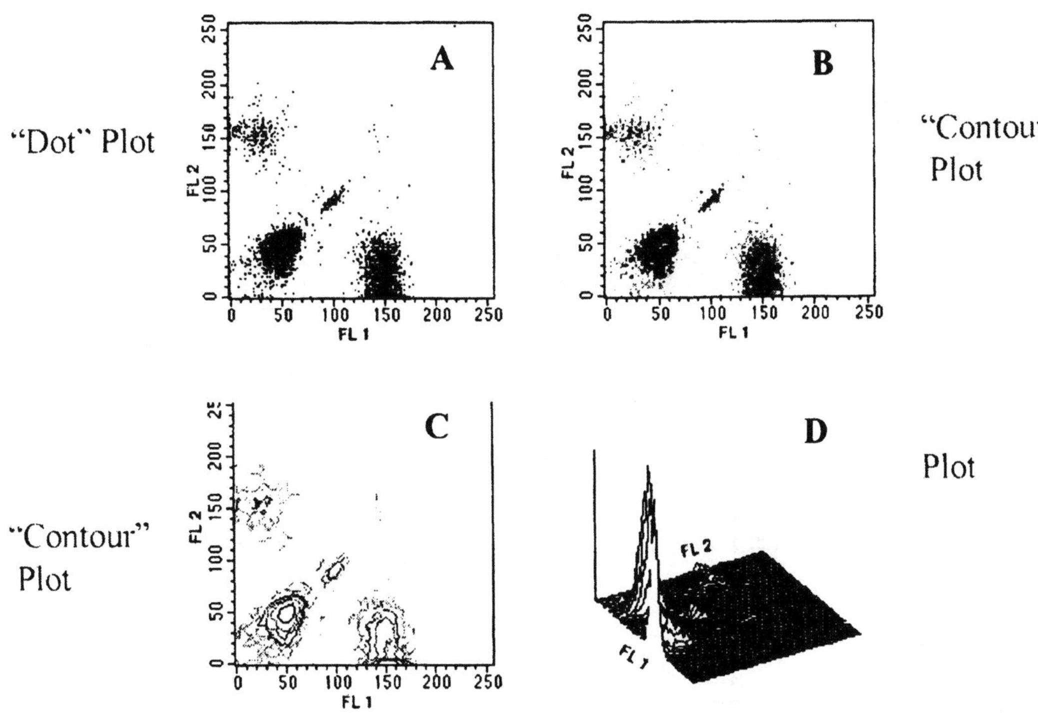

Figure 12. Four different two-parameter data displays of the same data.

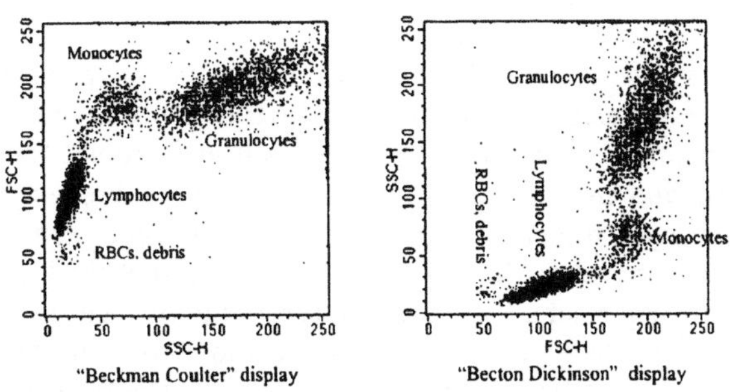

Figure 13. Light scattering properties reflect intrinsic physical differences among cells.

internal cell structure. Figure 13 is a classical example of the resolution achieved using light scatter analysis.

This data was accumulated using a sample of whole blood from which most of the red blood cells were removed. The three major populations of white blood cells are resolved with minimal overlap in their populations. This use of light scatter has proven to be enormously useful for clinical analysis of non-solid tumors when combined with fluorescent antibody staining. As illustrated in Figure 14.

In this instance, light scatter analysis was combined with fluorescein labeled antibodies to resolve lymphocytes, monocytes and polymorphonuclear leukocytes. Any of these populations could be sorted for functional studies with greater than 95% purity.

2.8. *Cell sorting*

Electrostatic cell sorting allows recovery of cells or chromosomes with predefined properties. Any parameter or combination of parameters available for use in a flow cytometer can be used as a basis for selection. Cell sorters are the method of choice for identifying and purifying subpopulations of cells or chromosomes for cloning purposes or construction of chromosome specific libraries, respectively. The principles of flow sorting are illustrated in Figure 15 and have been described in great detail in several publications. The principles of flow sorting are based on electrostatic deflection of charged droplets, some of which contain cells. Precise determination of the amount of time elapsed between cell sensing and analysis and the arrival of the cell at the point in space where droplets are formed allows charging the precise droplet containing a cell with a predetermined property. A charging voltage is applied in a precise fashion to coincide with the formation of a droplet containing a cell of interest. The process of applying a charging pulse and the arrival of a cell at the break-off point is critical. To accommodate some small amount of uncertainty in timing, three droplets are sometimes sorted, one droplet preceding the droplet believed to contain the cell and one droplet following the intended droplet. After the droplet is given a charge, the droplet(s) is deflected as it passes between two high voltage deflection plates. Sort purity is often an important criteria. Purity and yield is determined by a combination of how well the sorted population is resolved from other populations

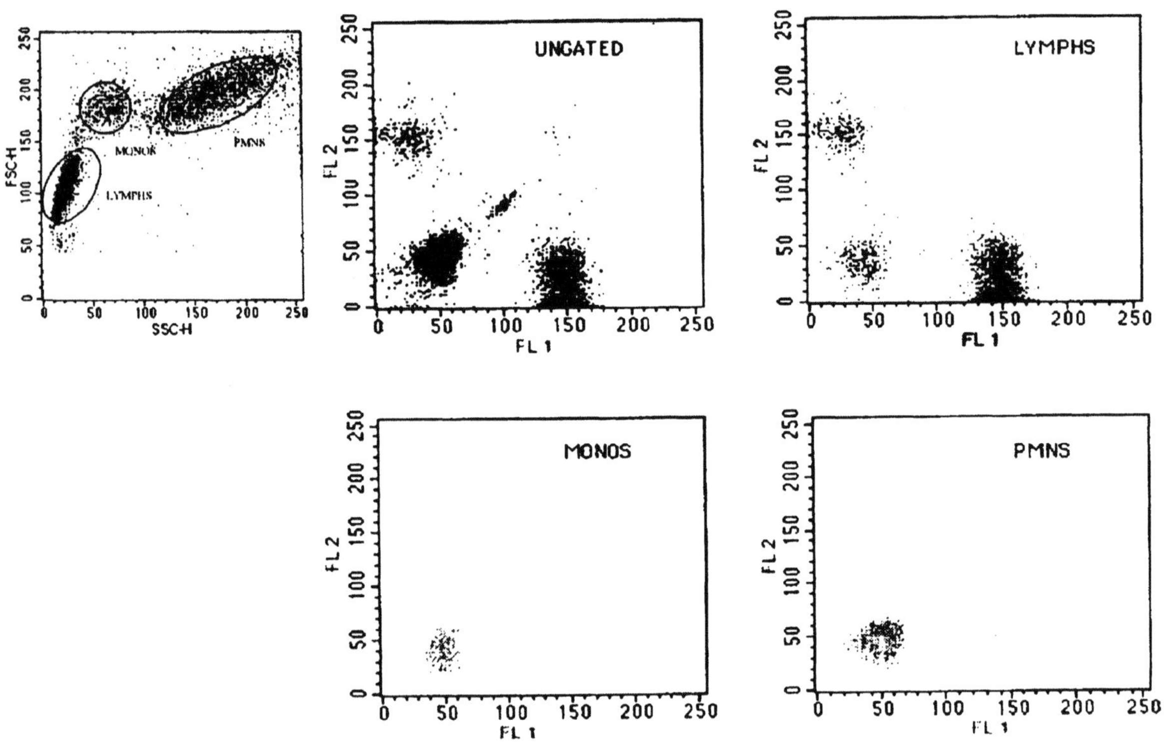

Figure 14. Light scatter gating for leukocytes.

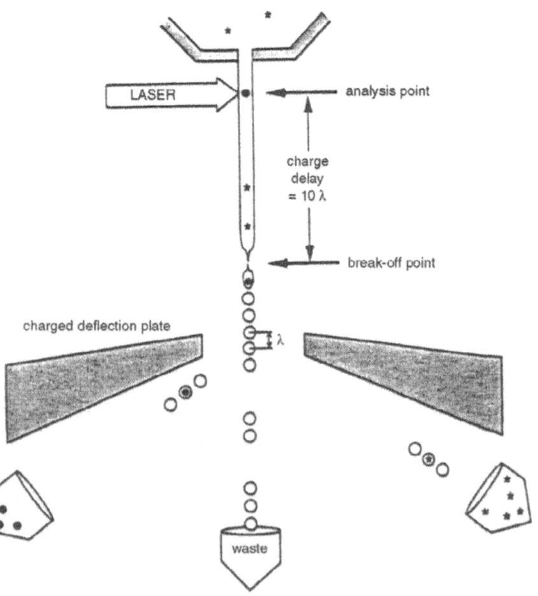

Figure 15. Cell sorting allows recovery of cells with predefined properties for functional characterization.

with similar characteristics, the rate at which cells are being analyzed (stream diameter), and several instrumental characteristics. For high purity sorting the ratio of droplets occupied to the number that are empty is important. A droplet frequency of 40 kHz together with an analysis rate of 1 kHz (1,000 cells/sec) provides, statistically, 39 empty droplets between every occupied droplet. Bubbles, debris, salt crystals, pressure fluctuations, and any movement of the droplet break-off point will contribute to small instabilities that disrupt the stream and droplet formation and lead to low rates of recovery and reduced purity. Droplet delay (the time between event sensing and arrival at the break-off point) is established by sorting a small number of representative cells or microspheres onto a microscope slide, counting the recovered fraction in a microscope and comparing the recovery to the number of events counted by the electronics.

High speed, high pressure flow sorting requires higher operating pressure (40 psi versus 12 psi), higher droplet frequency (65 kHz versus 32 kHz), and electronics capable of handling analysis rates of 10,000–20,000 events per second versus 2,000 events per second. The sort rate for all sorters depends on the fraction of a sample to be sorted and not the analysis rate. High speed, high pressure sorting presents challenging requirements when sorting certain types of cells, especially if viable cells are required. To take full advantage of high-speed high pressure sorting, particle concentration must be increased. This is not always possible when dealing with viable cells or chromosomes in suspension.

3. Summary

Flow cytometry had its origins in the late 1960s with the development of the first cell sorter based on the principle of droplet deflection used by Varian

Corporation in their ink-writing oscillograph. Today the technology represents a $600–800M dollar a year industry when one includes the necessary reagents together with instrument sales. The ability to make rapid measurements on single cells and to sort cells or chromosomes has revolutionized the study of heterogeneous populations. The most salient characteristics of the technology are the analysis rate (from 100's/second to 10,000/second), particle analysis time (msec to µsec), amount of data acquired (up to 50 parameters per cell), spatial resolution (zero dimensional), and the ability to physically separate cells and chromosomes for further study and/or applications. New flow cytometric technologies continue to be developed, as do applications that have a significant impact on our understanding of molecular cell biology and make dramatic contributions to clinical medicine.

4. Further reading

1. Melamed, M.R., Lindmo, T. and Mendelsohn, M. L. (eds) (1990) Flow Cytometry and Sorting, Second Edition, Wiley-Liss, New York.
2. Shapiro, H. M. (1988) Practical Flow Cytometry, Second Edition, Alan Liss, New York.
3. Omerod, M. G. (ed.) (1990) Flow Cytometry, A Practical Approach, Oxford University Press.
4. Gray, J. W. (ed.) (1989) Flow Cytogenetics, Academic Press Limited, San Diego, CA.

Acknowledgments

Appreciation is given to Dr's. Katharine Murihead and Alice Givan for their valuable contributions of many of the figures used in this manuscript. Appreciation is also expressed to all the outstanding members of the National Flow Cytometry Resource at Los Alamos National Laboratory without whose valuable contributions much of this work would not have flourished. This work was conducted under the auspices of the Department of Energy and with support from the National Institutes of Health, Division of Research Resources, Grant RR01315.

Methods in Cell Science **24**: 11–18 (2002).

High resolution flow cytometric analysis of electronic nuclear volume and DNA content in normal and abnormal human tissue

Richard A. Thomas[1], Awtar Krishan[2] & Michael Brochu[1]
[1] *NPE Systems, Inc., Miami, Florida;* [2] *Division of Experimental Therapeutics, Department of Radiation Oncology, University of Miami Medical School, Miami, Florida*

Abstract. Background: The NPE Analyzer® flow cytometer can simultaneously analyze the electronic nuclear volume (ENV) and DNA content of cells. This study describes the schematics, resolution, reproducibility, and sensitivity of biological standards analyzed on this unit.
Methods: Calibrated beads and biological standards (lymphocytes, trout erythrocytes [TRBC], calf thymocytes, and tumor cells) were analyzed for ENV versus DNA content. Parallel data (forward scatter versus DNA) from a conventional flow cytometer were obtained.
Results: ENV linearity studies yielded an R value of 0.999. TRBC had a coefficient of variation (CV) of 1.18 ± 0.13. DNA indexes as low as 1.02 were detectable. DNA content of lymphocytes from 42 females was 1.9% greater than that for 60 males, with a noninstrumental variability in total DNA content of 0.5%. The ENV/DNA ratio was constant in 15 normal human tissue samples, but differed in the four animal species tested. The ENV/DNA ratio for a hypodiploid breast carcinoma was 2.3 times greater than that for normal breast tissue.
Conclusions: The high-resolution ENV versus DNA analyses are highly reliable, sensitive, and can be used for the detection of near-diploid tumor cells that are difficult to identify with conventional cytometers. ENV/DNA ratio may be a useful parameter for detection of aneuploid populations.

Key words: DNA, Electronic nuclear volume, Flow cytometry, Forward scatter, Inflight cytometer, Nuclear packing efficiency

1. Introduction

Flow cytometry is an important method for rapid analysis of DNA content and phenotypic markers in cells. Several DNA binding/intercalating fluorochromes (e.g. propidium iodide, DAPI) are used to directly stain nuclei and detect aneuploidy and determine cell cycle distribution. However, due to technical limitations, it is often difficult to detect cells with low levels of aneuploidy (DNA index < 1.1) or detect sub-populations that cannot be distinguished on the basis of their DNA content and/or forward light scatter alone.

In most of the flow cytometric studies, light scatter (forward angle or 90 degree) is used for gating (elimination of debris) and selection of sub-populations for DNA and phenotype analysis. Although light scatter is related to the size of a cell, it is not a good measure of volume. In contrast, Coulter principle based on the measurement of impedance is an ideal tool for determination of volume in a flow system.

Distortions of the nuclear shape, size, and staining variations have been recognized as important diagnostic markers for the identification and classification of human tumors. The subjective study of these morphological variables is used in several histological grading systems used in tumor pathology. Simultaneous high-resolution measurement of nuclear volume and DNA content can provide new quantitative information that correlates with morphological parameters used by pathologists for tumor grading.

2. Materials

NIM-II DAPI, a single step combination of an isotonic Nuclear Isolation Medium containing 10 μg/ml of 4′, 6-diamidino-2-phenylindole (DAPI) dihydrochloride (Sigma Chemical Co., St. Louis, MO) at pH 7.2, and 0.6% NP-40 (Accurate Scientific and Chemical Co., Hicksville, N.Y.), was used for staining of the nuclear DNA [5].

2.1. *Biological standards*

Trout (*Salmo gairdnerii irideus*) red blood cells (TRBC) obtained from the U.S. Fish Hatchery, Erwin, Tenn. [2–4], were prepared at a concentration of 3×10^7 cells per ml (NPE Systems, Miami, FL) and were used as a biological DNA standard. A single lot was used during the course of this study.

Eighty-five human peripheral blood (PBL) samples (50 male and 35 female) were supplied by the Italian Red Cross, Rome, Italy. All samples were from healthy volunteer donors. Twelve samples (7 male and 5 female) came from volunteers in Miami, FL. In addition, 5 samples (3 male and 2 female), screened at the Department of Transfusion Medicine, National Institutes of Health, Bethesda, Maryland as normal human blood, were added to this study. In total, 102 (60 male and 42 female) samples were used with ages ranging 16–56.

3. Procedures

In the present study, we show data on simultaneous measurement of ENV vs. DAPI/DNA fluorescence of nuclei from human lymphocytes, cell lines, and human solid tumors, both fresh/frozen and paraffinized, analyzed on the NPE Flow Cytometer

Figure 1 shows the optical and electronics schematic of the NPE Flow Cytometer. One of the unique and innovative engineering features of this unit is that Electronic Volume (Coulter Volume) measurement is made simultaneously with fluorescence measurement in the same spatial location. This removes the ambiguity that may be caused by measurement of these two parameters at different spatial locations. The ability to perform simultaneous electronic volume and high-resolution fluorescence measurements is made possible by the unique geometry of the flow cell shown in Figure 1. The flow cell, 70 μM in length, has an equilateral triangular cross-section of 70 μM per side. The inlet and outlet chambers also have equilateral triangular cross-sections. Their dimensions start at 2 cm per side, and decrease over 0.7 cm to 70 μM per side. The triangular flow cell geometry produces large hydrodynamic forces that focus the sample stream to a small region of the 3 dimensional space than is possible with circular, rectangular or square flow cell geometry [1]. This highly reproducible sample flow pathway makes it possible to use an oil immersion objective with high numerical aperture (N.A.) resulting in greater photon collection efficiency. The stream stability also makes the pathway through the excitation beam more reproducible. These factors in combination result in lower intensity requirements for the light source, greater signal to noise ratio and higher resolution.

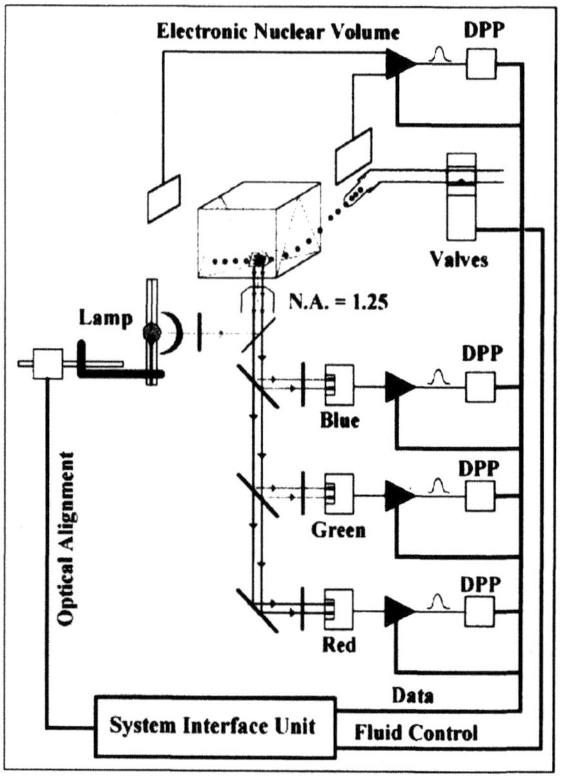

Figure 1. Optical and Electronics Schematic of the NPE Analyzer, from Thomas et al. Cytometry 43: 2–11, 2001 (22).

The micro-objective used has a 1.25 N.A corresponding to a light collection angle of 120 degrees for the fluorescence emission. This approximates the maximum theoretical light gathering N.A. from the water/glass interface of the flow cell. The light gathering efficiency allows the use of a solid-state photo detector instead of a conventional photo multiplier tube for DNA measurement and a standard mercury arc lamp excitation light source instead of a laser or cadmium doped arc lamp. The instrument employs epi-illumination collection optics [18] with a dichroic mirror to reflect the excitation beam and allow passage of the fluorescence emission. The flow cell geometry allows for simultaneous collection of both optical and electronic volume measurements. The 100-watt stabilized mercury arc lamp has a 30 μM spot size selected from the most uniform region of the arc. This large spot size provides uniform illumination for analysis of a wide range of particle sizes, while providing sufficient photon flux density for fluorescent antibody measurements.

The entire optical system alignment is achieved by computer controlled stepping motors and normally takes 3–4 min to achieve optimal results. Optical alignment is not necessary on a day-to-day basis, but is performed when the CV of the DAPI stained TRBC used for calibration, became greater than 1.4%, or after lamp replacement.

The fluidic system consists of 17 computer-controlled valves programmed to work through the computer interface to automatically perform all the sample handling and flushing operations. Fluid transfer is accomplished by a vacuum pressure differential of 10 mm of Hg.

Digital Pulse Processing (DPP) is used to determine the peak value of the DNA and ECV pulses. The DPP electronics package was designed to provide rapid, low noise peak detection by sampling the pulse every 200 nanoseconds. The maximum digital value of the approximately 200 samples from every pulse is the digital peak value. DPP has eliminated the variation in peak sampling error with the pulse size, which occurs in analog peak detection systems. Small variations in the peak location of the internal DNA standard (TRBC) are detected by the DPP electronics and used to servo the solid-state photo detector output. This real time servo reduces instrument drift, and improves the absolute calibration of the system in terms of TRBC equivalents of DNA.

A two ml suspension of the TRBC was prepared daily by mixing 150 µl of the TRBC solution in 2 ml of NIM II – DAPI resulting in a stock solution of stained TRBC nuclei at a concentration of approximately 2×10^6 nuclei/ml. The stained TRBC nuclei were filtered through a 35 µM polyethylene frit filter (NPE Systems, Miami, FL) and refrigerated at 4 °C until used. If the CV of the TRBC standard was greater than 1.4%, the automatic optical alignment program was run. After successful completion of the alignment process, the instrument gain was calibrated by placing the TRBC DNA peak in channel 50 of the 256-channel pulse height analyzer.

Samples were prepared by adding 100 µl of fresh whole blood to 1 ml of NIM – DAPI solution. Two human PBL samples from the Miami Florida group were prepared by adding 100 µl of fresh whole blood to 1 ml of NIM – DAPI solution. One sample was from a male and the other from a female volunteer. Five consecutive analyses were run on each sample, reinjecting the sample between analyses. A second set of samples was prepared and an additional 5 consecutive analysis were run on each sample for a total of 10 analyses per sample performed in one hr. Following these analyses, the two blood samples from the male and the female volunteer were mixed in equal amounts, stained and analyzed as described above.

3.1. *Sample preparation*

Tumor samples were prepared by mincing 1 cubic mm of fresh or frozen tissue with a surgical scalpel in 1 ml of NIM – DAPI or propidium iodide (PI) solution (50 µg/ml) containing 1% citrate and 0.1% NP-40 (Accurate Scientific and Chemical Co., Hicksville, N.Y.).

All samples were filtered through a 35 µM polyethylene frit filter (NPE Systems, Miami, FL) and refrigerated at 4 °C until analyzed. Approximately 20 µl of the stained TRBC stock suspension was added to each sample as an internal standard for calibrating the sample in TRBC equivalent pg of DNA. After analysis of every 100 TRBC nuclei, the photodetector output gain was automatically adjusted to keep the TRBC DNA peak in channel (50 ± 0.25), thus compensating in real time for any instrument drift. With this calibration, each channel equaled 0.047 pg of TRBC equivalent DNA. The specimens were analyzed at a flow rate of 60–120 nuclei/second and 15,000 events were collected in a FSC 2.0 Standard list mode file.

3.2. *Data analysis*

Cell cycle analysis of the DNA distribution histograms was performed by Modfit version 5.11 (Verity Software House Inc., Topsham, Maine, USA). Multi-parameter graphics were prepared using WinMDI (version 2.7) downloaded from the World Wide Web site, http://facs.scripps.edu. Sigma Plot 4.01 software (SPSS Inc. Chicago, IL, USA) was used for statistical analysis. All values are expressed as mean values ± 1 Standard Deviation.

3.3. *Human peripheral blood leukocytes*

Figure 2 is a plot of TRBC equivalent pg of DNA for each of the 102 human PBL samples analyzed. The error bars are one standard deviation for each of the male and female mean value data sets. The DNA content per nucleus for the female samples was 3.77 ± 0.03 TRBC equivalent pg of DNA with a mean CV of the DNA histograms of 2.1 ± 0.15%. The DNA content per nucleus for the male samples was 3.70 ± 0.025 TRBC equivalent pg of DNA with a CV of 2.0 ± 0.2%.

The DNA content value in TRBC equivalent pg of DNA for the female and male samples from Rome, Italy was 3.77 ± 0.023 (n = 35) and 3.70 ± 0.025 (n = 50), respectively. The value for the female and male samples from Miami, Florida was 3.79 ± 0.065 (n = 5) and 3.70 ± 0.027 (n = 7), respectively. The value for the female and male samples from Bethesda, MD was 3.75 ± 0.026 (n = 2) and 3.69 ± 0.014 (n = 3), respectively.

Figure 3 is a plot of 10 repetitions of one male and one female PBL specimen analyzed to test stability and reproducibility of the sample preparation and instrument system. The analyses were run sequentially, first the male then the female, alternating back and forth using the same two samples over and over. After 5 pairs of analyses, a new set of samples were prepared and 5 additional pairs of analyses were run using the same technique. Sample 1 was the first pair run and sample 10 was the last pair run. All twenty samples were run during the course of 1 h. As shown in Figure 5, DNA content of the female (open triangles) and the male (solid triangles) PBL's were 3.75 ± 0.01 and 3.68 ± 0.009 TRBC equivalent pg of DNA, respectively. The CV of the DNA histograms for the female and male PBL's was 1.4 ± 0.3% and 1.6 ± 0.2%, respectively. The data from the analysis of the duplicate sample were similar to the first.

Figure 4 shows a DNA histogram of a 50/50 mixture of the male and female PBLs used to produce the data in Figure 5. The population at channel 100

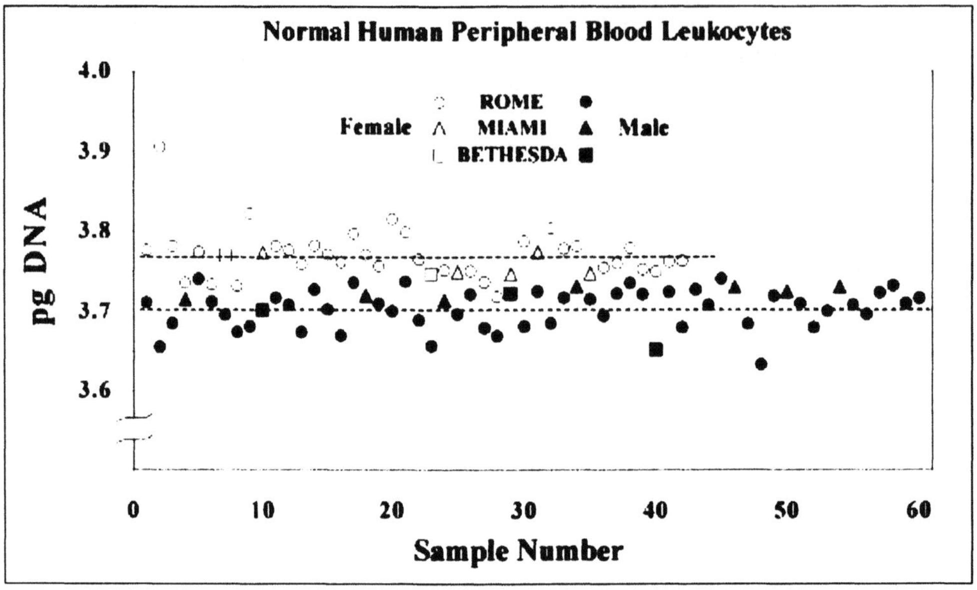

Figure 2. Normal Human Peripheral Blood Leukocytes, from Thomas et al. Cytometry 43: 2–11, 2001 (22).

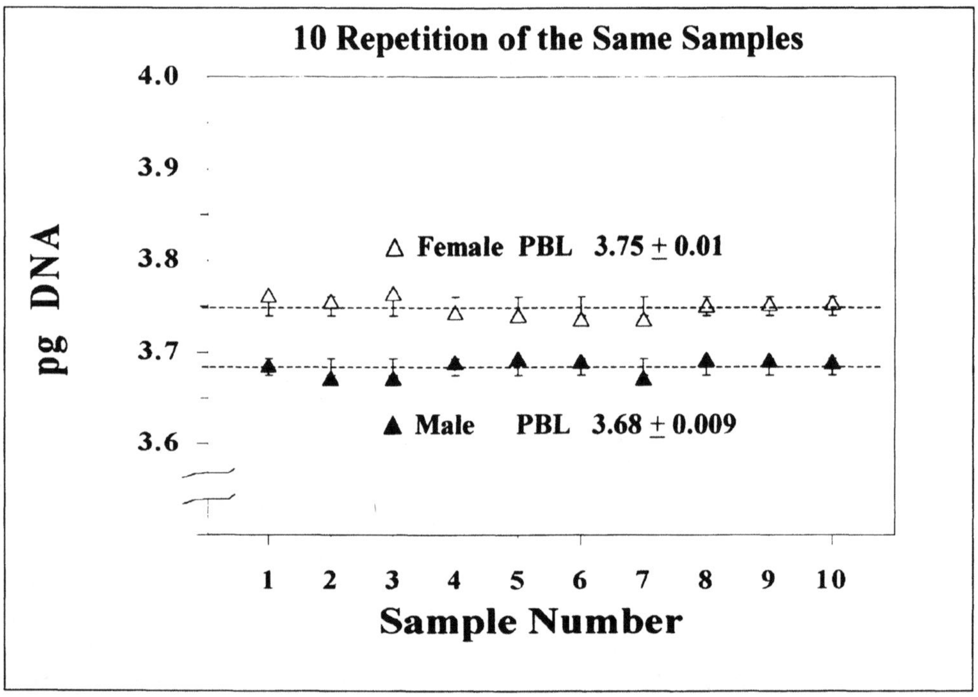

Figure 3. Stability and Reproducibility of the Sample Preparation and Instrument System, from Thomas et al. Cytometry 43: 2–11, 2001 (22).

is the TRBC standard with a CV of 1.3%. There are two distinct populations centered about channel 156. The male PBL peak is located at channel 155 corresponding to 3.674 pg DNA and the female PLB peak is located at channel 158 corresponding to 3.745 pg DNA.

4. Results and discussion

4.1. *Identification of near-diploid tumor cells*

In some tumors most cytometer can readily identify an aneuploid tumor populations by the presence of nuclei with DNA index greater than 1.1 but have broad diploid peaks that are suggestive of a possible near diploid population. Figure 5:A,B is an ovarian tumor which exemplifies this type of histogram. The

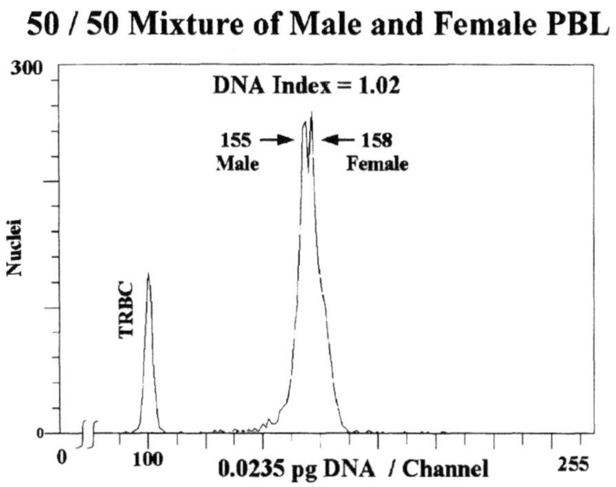

Figure 4. 50/50 Mixture of Male and Female PBLs, from Thomas et al. Cytometry 43: 2–11, 2001 (22).

TRBC internal standard peak in this histogram has a narrow CV while the G_0/G_1 peak is broad suggesting the presence of aneuploid sub-population (indicated by the arrow), which could not be resolved as a distinct peak. As shown in Figure 5B (ENV vs. DNA content), the iso-contour plot shows that the nuclei in the near-diploid population have a much larger range of ENV (arrow) than that of the normal diploid G_0/G_1 nuclei. The NPE line for normal human diploid nuclei is shown as the black line passing through the origin in this Figure.

Examples of NPE analysis of normal human diploid nuclei are shown in Figures 6A and 7A, and also may be found in Figure 7B of the accompanying paper [10]. There are two distinct aneuploid populations (aneuploid 1 and aneuploid 2) identifiable by their NPE signature in this tumor, one with a DNA Index (the ratio of the aneuploid DNA peak channel number to the diploid DNA peak channel number (DI)) of 1.05 and the other with DI of 1.45.

In some tumor populations the DNA distribution histogram may show a small shoulder on the upper side of the main G_0/G_1 peak indicating the possible presence of a near-diploid/aneuploid population. The DNA histogram of a prostate tumor shown in Figure 5C has this characteristic. In Figure 5D (the iso-contour plot of ENV vs. DNA) the nuclei associated with the shoulder on the DNA histogram (arrow) have elevated ENV displaying a NPE signature characteristic of tumor cells. The line connecting the G_0/G_1 (2C) and the G_2/M (4C) populations is the Nuclear Packing Efficiency (NPE) line for normal human cells [10].

Using DNA histograms alone (Figure 5A,C) it is difficult to identify the near-diploid/aneuploid populations. However, in the two-parameter iso-contour plots (Figure 5B, D), these populations are identifiable as having nuclei with 3 to 4-fold greater ENV and a DI of 1.05. The distinctive NPE signature confirms the presence of aneuploidy in these tumors.

4.2. *Identification of metastatic tumor cells*

The use of ENV vs. DNA can often be used to determine the presence of aneuploid tumor cells in

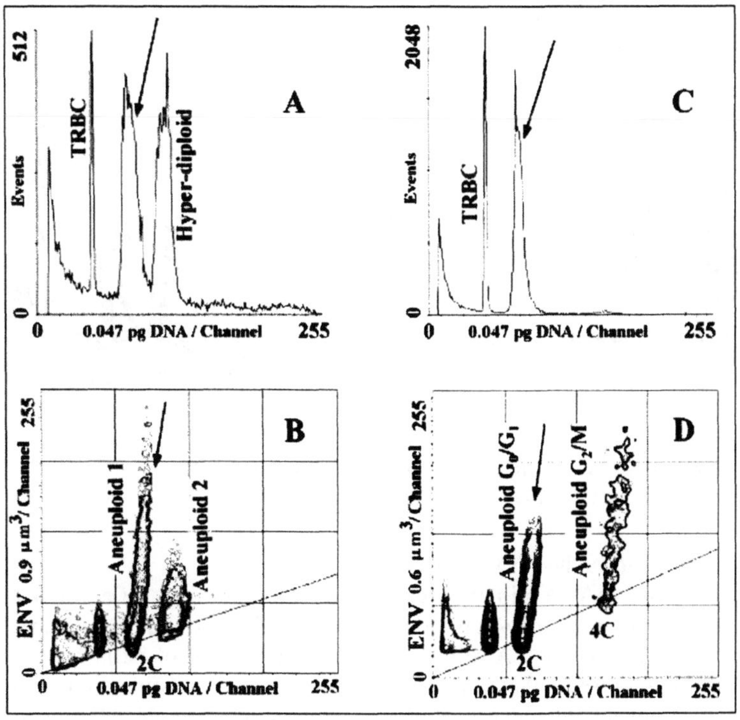

Figure 5. A&B: Histogram and Iso-Contour Plot of an Ovarian Tumor; C&D: Histogram and Iso-Contour Plot of a Prostate Tumor, from Thomas et al. Cytometry 43: 2–11, 2001 (22).

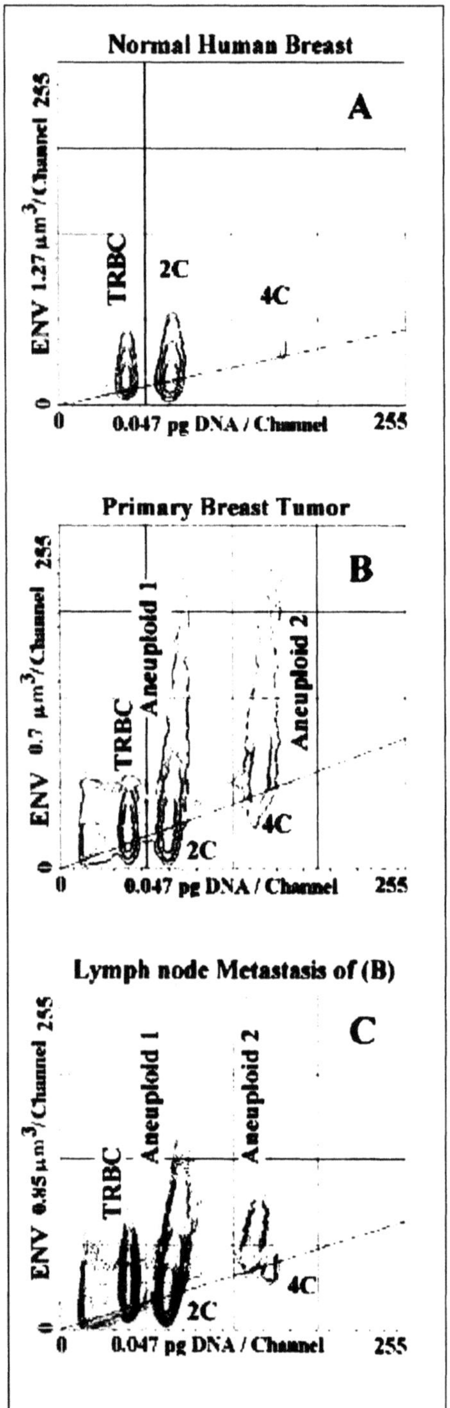

Figure 6. ENV vs. DNA Content of Normal Breast Tissue, from Thomas et al. Cytometry 43: 2–11, 2001 (22).

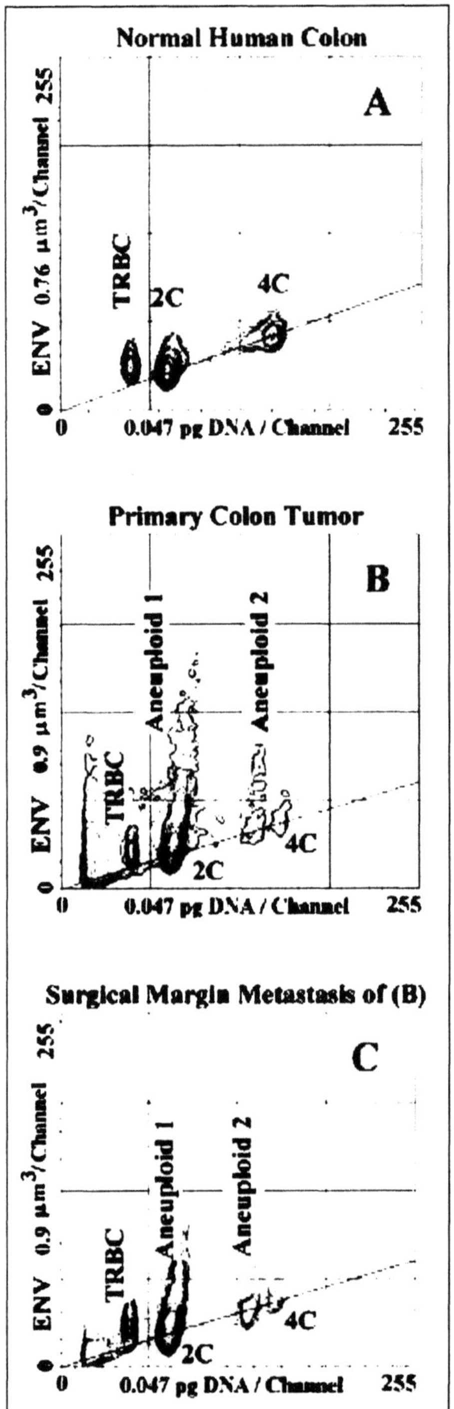

Figure 7. ENV vs. DNA Content of Normal Human Colon Tissue, from Thomas et al. Cytometry 43: 2–11, 2001 (22).

metastatic tumors. Figure 6 shows the ENV vs. DNA content of normal breast tissue (6A), a primary breast tumor (6B), and a lymph node with histo-pathologically confirmed tumor infiltration (6C). A reference line drawn across the plots represents the NPE value of normal nuclei and connects the center of the diploid G_0/G_1 (2C) and G_2/M peaks (4C) [10].

Nuclei from the normal breast tissue (Figure 6A) are characterized by a small ENV signal (confined within the 1st quadrant of the Y-axis) and a pattern similar to that of cells from normal human breast, colon, small intestine, thyroid, lymph node tissue and lymphocyte samples [10]. Nuclei from these normal tissues had similar low ENV with a mean volume of 21 ± 0.5 μm^3. In contrast the ENV signal of the aneuploid nuclei (DI of 1.06) from the primary breast tumor (Figure 6B, aneuploid 1) was spread over four quadrants of the Y-axis. Figure 6B also shows the presence of a near-tetraploid population (aneuploid 2) with high ENV and DI of 1.75. The ENV pattern seen in this tumor was characteristic of several other poorly differentiated breast tumors studied (data not shown).

Figure 6C shows nuclei isolated from a lymph node of the node positive patient whose primary tumor is shown in 6B. The lymph node of this patient was histopathologically confirmed to contain tumor cells. A predominant population of diploid nuclei and a smaller population of nuclei with G_2/M DNA content lie along the normal tissue NPE line. Nuclei from normal human lymph nodes (as shown in Figure 8B of the accompanying paper, 10) have in general a small ENV signal confined within the 1st quadrant of the Y-axis. The diploid G_0/G_1 and G_2/M nuclei in this lymph node are on the NPE line and stand distinct from the aneuploid 1 and aneuploid 2 nuclei with larger ENV. A comparison of the aneuploid populations of the primary tumor (Figure 6B) with the aneuploid populations in the lymph node (Figure 6C) shows that the distinct NPE pattern seen in the primary tumor was also seen in the lymph node. The population marked Aneuploid 2 in Figure 6C (representing 3% of the total population) illustrates the sensitivity of ENV vs. DNA method for distinguishing between normal diploid G_2/M and near-tetraploid tumor cells and detecting small number of aneuploid nuclei in a tumors.

ENV vs. DNA analysis can also be used for detecting the presence of tumor cells in surgical margins and distant metastasis of a tumor. Figure 7 shows a comparison of ENV/DNA content in nuclei from normal colon (Figure 7A), a primary adenocarcinoma of the colon (Figure 7B) and in distant metastasis of this tumor (Figure 7C).

The population of diploid nuclei is characterized by a small ENV signal (confined within the 1st quadrant of the Y-axis) with a prominent G_2/M (7C) population falling on the normal tissue NPE line. Figure 7B of the primary tumor shows a near-diploid/aneuploid population (Aneuploid 1) (DI of 1.09) with very large nuclei and a second smaller near-tetraploid population (Aneuploid 2) with DI of 1.75, clearly distinguished by its ENV. Figure 7C is of tissue from the end portion of the surgical resection of the patient's primary tumor shown in 7B. A comparison of the aneuploid populations of the primary tumor (Figure 7B) with the aneuploid populations in the colon resection (Figure 7C) shows that the distinct NPE pattern seen in the primary tumor was also seen in secondary site. The approximately 3% tumor cells in this resection were not detected in the initial histological examination but subsequent re-examination confirmed the presence of tumor cells at this site.

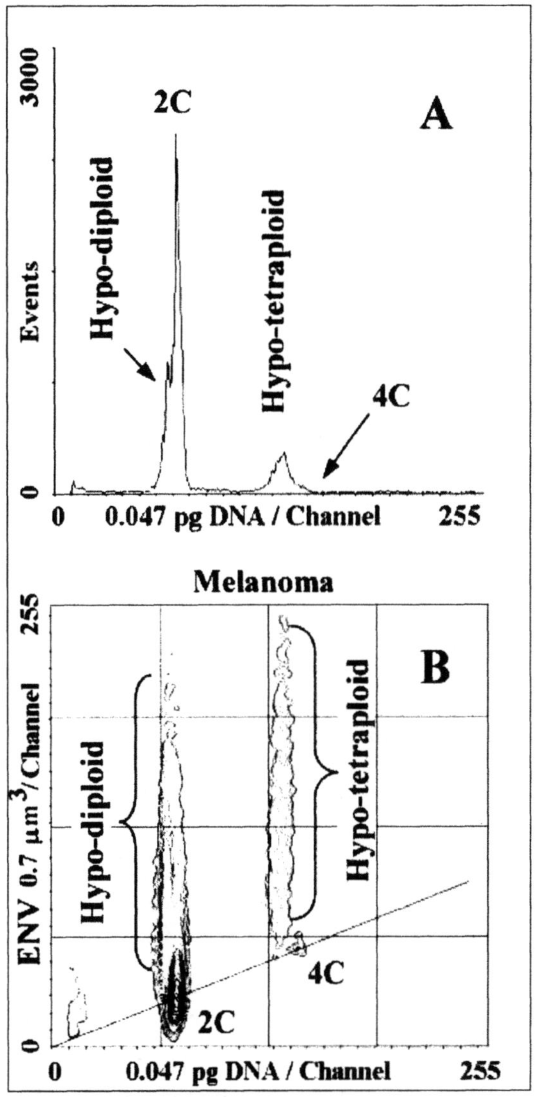

Figure 8. Differentiating Hypo-diploid from Hyper-diploid on a Melanoma Sample, from Thomas et al. Cytometry 43: 2–11, 2001 (22).

4.3. *Differentiating hypo-diploid from hyper-diploid*

ENV vs. DNA analysis can often facilitate discrimination between hypo-diploid and diploid cells in a population. Figure 8A shows DNA histogram of a melanoma with a hypo-diploid shoulder (arrow). Figure 8B shows that the nuclei with the larger ENV are from the population containing hypo-diploid DNA content with DI of 0.94. The large population of hypo-tetraploid nuclei with elevated ENV (Figure 8B) further confirms this. The normal tissue NPE line passes through the center of the diploid (2C) and the small diploid G_2/M (4C) population.

Acknowledgments

The NPE Flow Cytometer referenced herein is based on an analyzer previously identified as the NASA/American Cancer Society (ACS) Flow Cytometer, developed by Ratcom, Inc. NPE Systems, Inc. has obtained license to this product and technology from Ratcom, Inc., and has named the product the NPE Analyzer®.

This article is excerpted from two original articles that first appeared in Cytometry, Volume 43, Issue 1, 2001, published 2001 Wiley-Liss, Inc.; Richard Thomas article Pages 2–11, and Awtar Krishan article Pages 16–22.

Supported by the American Cancer Society, Florida Div. Grant (M91-00440) and the National Aeronautics and Space Administration contract (NAS 9-18379).

References

1. Krishan A, Wen JH, Thomas RA, Sridhar KS, Smith WF. NASA/American Cancer Society high-resolution flow cytometry project – II: Multiparametric analysis of DNA content and electronic nuclear volume in human solid tumors (manuscript submitted).
2. Thomas R, Eggleston W (1987). Flow cytometry. U.S. Patent No. 4,673,288.
3. Ploem JS (1967). The use of a vertical illuminator with interchangeable dichroic mirrors for fluorescence microscopy with incidental light. Z Wis Mikrosk 68: 129–142.
4. Thornthwaite JT, Sugarbaker EV, Temple WJ (1980). Preparation of tissues for DNA flow cytometric analysis. Cytometry 1: 229–237.
5. Fasman GD (1976). Selected Data for Molecular Biology In: The Chemical Rubber Co. (CRC) Handbook of Biochemistry and Molecular Biology Nucleic Acids Volume II, 3rd Edition, p 296.
6. Knobloch A, Vendrely C, Vendrly R (1957). The amount of deoxyribonucleic acid in a single trout sperm. Biochim Biophys Acta 24: 201.
7. Lee GM, Thornthwaite JT, Rasch EM (1984). Picogram per cell determination of DNA by flow cytofluorometry. Anal Biochem 137: 221–226.
8. Deloukas P, Schuler GD, Gyapay G, Beasley EM, Soderlund C, Rodriguez-Tome P, Hui L, Matise TC, McKusick KB, Beckmann JS, Bentolila S, Bihoreau M, Birren BB, Browne J, Butler A, Castle AB, Chiannilkulchai N, Clee C, Day PJ, Dehejia A, Dibling T, Drouot N, Duprat S, Fizames C, Bentley DR et al. (1998). A physical map of 30,000 human genes. Science 282: 744–746.
9. Harris P, Boyd E, Young BD, Ferguson-Smith MA (1986). Determination of the DNA content of human chromosomes by flow cytometry. Cytogenet Cell Genet 41: 14–21.
10. Morton NE (1991). Parameters of the human genome. Proc Natl Acad Sci USA 88: 7474–7476.
11. Mullaney PF, Crowell JM, Salzman GC, Martin JC, Hiebert RD, Goad CA (1976). Pulse-height light-scatter distributions using flow-systems instrumentation. J Histochem Cytochem 24: 298–304.
12. Leary JF, Todd P, Wood JC, Jett JH (1979). Laser flow cytometric light scatter and fluorescence pulse width and pulse rise-time sizing of mammalian cells, J Histochem Cytochem 27: 315–320.
13. Mullaney PF, Van Dilla MA, Coulter JR, Dean PN (1969). Cell sizing: a light scattering photometer for rapid volume determination, Rev Sci Instrum 40: 1029–1032.
14. Stal O, Hatschek T (1988). A rapid system for static cytofluorometry enabling the simultaneous determination of nuclear size and DNA content. Pathol Res Pract 183: 329–335.
15. Grover NB, Naaman J, Ben-Sasson S, Dolianski F (1969). Electrical sizing of particles I. Biophys J 9: 1398–1414.
16. Thomas RA, Cameron BF (1974). Electronic cell volume analysis utilizing the AMAC II transducer. J Histochem Cytochem 22: 626–641.
17. Petersen SE (1986). Accuracy and reliability of flow cytometric DNA analysis using a simple, one-step ethidium bromide staining protocol. Cytometry 7: 301–306.
18. Cram LS, Lehman JM (1977). Flow microfluorometric DNA content measurements of tissue culture cells and peripheral lymphocytes. Human Genetics 37: 201–206.
19. Hoehn H, Johnston P, Callis J (1977). Flow-cytogenetics: sources of DNA content variation among euploid individuals. Cytogenet Cell Genet 19: 94–107.
20. Redkar A, Krishan A (1999). Flow cytometric analysis of estrogen, progesterone receptor expression and DNA content in formalin fixed, paraffin embedded human breast tumors. Comm Clin Cytometry 38: 61–69.
21. Krishan A, Oppenheimer A, You W, Dubbins R, Sharma D, Lokeshwar B (2000). Flow cytometric analysis of androgen receptor expression in human prostate tumors and benign tissues. Clin Cancer Res 6: 1922–1930.
22. Thomas A, Krishan A, Robinson D, Sams C and Costa F (2001). NASA/American Cancer Society High-Resolution Flow Cytometry Project-I. Cytometry 43: 2–11.
23. Krishan A, Wen J, Thomas R, Sridhar K, Smith W (2001). NASA/American Cancer Society High-Resolution Flow Cytometry Project-III. Multi-parametric Analysis of DNA Content and Electronic Nuclear Volume in Human Solid Tumors. Cytometry 43: 16–22.

Author for correspondence: Richard A. Thomas, NPE Systems, Inc., 7620 SW 147 Court, Miami, FL 33193, USA
Phone: 305-382-3947; Fax: 954-566-6541
E-mail: ratcom@bellsouth.net

Methods in Cell Science 24: 19–25 (2002).

Analysis of heterogeneous red cell populations by flow cytometry

M. Nelson
Institute of Haematology, Royal Prince Alfred Hospital, Sydney, Australia

1. Introduction

The application of the technique of flow cytometry to the analysis of red blood cell (RBC) populations has not been as common as its use in the identification of various leukaemias and lymphomas [3, 11, 31] or in the field of platelet immunology [18, 29], although a report on the quantitation of the destruction of opsonized red cells appeared as early as 1980 [30]. The sensitivity and accuracy of the technique as applied to RBC populations has strongly recommended it for rare event analysis, and particularly in the detection of fetomaternal haemorrhage when the mother is Rh(D) negative and at grave risk of being sensitized by leakage across the placenta of fetal Rh(D) positive cells [10, 15, 22]. In many instances, a multi-transfused patient can be phenotyped through quantitation of blood group antigens. This can be difficult using traditional serological methods and economically impractical using DNA-based typing [5] but it is relatively simple by flow cytometry. The process of red cell engraftment following homologous bone marrow or peripheral stem cell transplant may be monitored provided there is a suitable mismatch of antigens. Continuing quantitation of the differences can give early warning of loss of the graft or confirmation of a stable chimaera [1, 4, 16]. Certain disease states may be characterised by the complete or partial loss of certain specific antigens, notably the glycosylphosphatidylinositol (GPI)-anchored antigens in paroxysmal nocturnal haemoglobinuria [21, 26], or changes due to the loss of A, B or H glycosyl transferases [14, 28] and even the loss of Rh(D) expression [6], in patients suffering various leukaemias.

These applications are of particular interest because flow cytometry techniques can quantitate antigenic mixtures with great accuracy. There are further advantages in that only a few hundred cells are needed for analysis, and this has been utilized in the determination of fetal blood group antigens on the RBC contained in fragments of chorion villus [23].

Not only can the antigenic composition of blood samples be investigated, but also the antibodies contained in maternal plasma or in the fetal circulation can be titrated and sub-classed, for example, using screening RBC of the appropriate phenotype [2, 7, 17]. Again, the volume of plasma required for such investigations can be very small and heelprick samples from neonates may be screened for residual maternal antibody. Although the tests directed towards antibody investigations can be of great value, they hardly fall into the category of 'heterogeneous populations of RBCs' and so will not be dealt with further here.

This article describes the methods and procedures for simple tests dealing with the quantitation of mixed red cell populations.

2. Materials and methods, surface antigens

In order to avoid repetition in giving details of companies supplying or manufacturing antibodies or pharmaceutical products of relevance for the analysis of RBC populations, Table 1 gives the contact details for these companies. The list is intended as a guide only, and may represent only a small number of the companies and products available.

2.1. *Reagents*

2.1.1. *Phosphate buffered saline (PBS)*

A general purpose buffer which is suitable for all applications of RBC in flow cytometry is made up as follows:

- 8.1 g sodium chloride in 900 ml distilled or deionized water (or 900 ml of physiological saline);
- 100 ml 0.15 **M** dipotassium hydrogen phosphate, pH 7.2

The commercially available Dulbecco's Buffer A tablets (Oxoid) in common use for leukocyte flow cytometry has been found to be not particularly suitable for RBCs because of high non-specific backgrounds, perhaps a consequence of the excipients incorporated in the tablets and their possible reaction with the highly glycosylated cell membrane. The imidazole buffer in use in the blood bank of this hospital (Lab Aids Pty. Ltd.) has also been found to be unsatisfactory. The addition of proteins such as bovine serum albumin to the PBS has been found to increase the number of background events without conferring additional stability on the cell suspensions, and so the use of such additives has been discontinued.

Table 1. Companies who manufacture or supply reagents used in the detection or quantitation of mixed RBC populations

Company	Address	Website
Beckman Coulter Inc.	Fullerton, CA, USA	www.beckman.coulter.com
Becton Dickinson Immunocytometry Systems	San Jose, CA, USA	www.bdfacs.com
BD PharMingen	San Diego, CA, USA	www.bdbiosciences.com
Biotec Laboratories Ltd.	Ipswich, Suffolk, UK	www.bioportfolio.com
Bio-Rad Laboratories	Hercules, CA, USA	www.bio-rad.com
Biotest AG	Dreieich, Germany	www.biotest.de
Caltag Laboratories	Burlingame, CA, USA	www.caltag.com
Calbiochem-Novabiochem Corp.	La Jolla, CA, USA	www.calbiochem.com
Chemicon Australia	Boronia, VIC, Australia	www.chemicon.com.au
CSL Ltd.	Parkville, VIC, Australia	www.csl.com.au
Gamma Biologicals Inc.	Houston, TX, USA	www.gammabio.com
Immucor Inc.	Norcross, GA, USA	www.immucor.com
Immunotech (Coulter)	Marseille, France	www.beckman.coulter.com
International Blood Group Reference Laboratory	Bristol, UK	www.bloodnet.nbs.nhs.uk
Jackson ImmunoResearch Laboratories Inc.	West Grove, PA, USA	www.jacksonimmuno.com
Lab Aids Pty. Ltd.	Narrabeen, NSW, Australia	–
Molecular Probes Inc.	Eugene, OR, USA	www.probes.com
Oxoid Ltd.	Basingstoke, Hamps. UK	www.oxoid.com
Serologicals Corp.	Livingston, Scotland, UK	www.serologicals.com
Serotec Inc.	Raleigh, NC, USA	www.serotec.com
Sigma-Aldrich Fine Chemicals	St Louis, MO, USA	www.sigma-aldrich.com
Wallac Inc.	Akron, OH, USA	www.wallac.com

2.2. *Antibodies*

2.2.1. *General comments*

The antibodies to blood group antigens that are suitable for use are of class IgG. Antibodies of class IgM tend to spontaneously agglutinate the red cells and can give rise to false positive results because of these aggregates. It is becoming increasingly difficult however to obtain IgG antibodies since most serological laboratories use monoclonal IgM antibodies for simplicity and speed in blood phenotyping and cross matching. Polyclonal antibody mixtures derived from pooled donor plasma can be treated with dithiothreitol (DTT) [27] to destroy the IgM component, or the classes of antibody may be separated through use of affinity chromatography (i.e. 'DEAE Affi-gel Blue'; Bio-Rad Labs.). Commercially available antibodies to blood group antigens A or B, M or N, are invariably IgM and so are not suitable for use in flow cytometry. Plasma from patients or donors of blood groups A, B or O (containing anti-B, anti-A or anti-A,B) can occasionally be used as reagents after judicious addition of 0.01 **M** DTT to destroy the IgM component, followed by titration of the treated plasma to find the appropriate dilution for use as a primary antibody.

2.2.2. *Directly labelled fluorescent-antibodies*

Directly conjugated human monoclonal IgG antibodies are available for only a limited number of antigens. Conjugated antibodies to Rh(D) have been developed as a consequence of their application in detecting fetomaternal haemorrhage in Rh(D) negative women bearing an Rh(D) positive baby. These antibodies include the fluorescein conjugate BRAD 3-FITC (International Blood Group Reference Laboratory) and Quanti-D, a phycoerythrin [PE]-conjugate (Chemicon Australia). Non-conjugated monoclonal anti-Rh(D) antibodies were originally used during the development of these tests [20] or as a biotinylated first antibody [22] followed by incubation with a fluorescent-labelled secondary antibody or streptavidin conjugate. Directly conjugated antibodies are also available for the glycosylphosphatylinositol (GPI)-linked antigens CD55, 58 and 59 which are expressed on the RBC membrane. The absence or reduced expression of these antigens on a proportion of the cells are characteristic of PNH. These antibodies are available from BD PharMingen, Beckman Coulter Inc. or Serotec Inc.

2.2.3. *Non-conjugated antibodies to blood group antigens*

Polyclonal blood grouping antibodies with a strong IgG component are available from a number of manufacturers such as Immucor Inc., Gamma Biologicals Inc., CSL Ltd., Biotest AG, Serologicals Corp. and Biotec Laboratories Ltd. These antibodies must be used in conjunction with a fluorescent-labelled secondary antibody, generally a polyclonal sheep or goat anti-human IgG and preferably as a $F(ab')_2$ fragment to eliminate nonspecific reactions. These antisera are available through the companies Chemicon Australia,

Jackson ImmunoResearch Laboratories Inc. and Sigma-Aldrich Fine Chemicals. Fluorescent-labelled mouse monoclonal antibodies to human IgG may be used also, obtainable from BD Pharmingen for example.

The primary antibodies of particular value in the analysis of mixed RBC populations are:

Rh phenotyping antisera [apart from Rh(D)] C, c E and e;
Duffy [Fy^a and Fy^b];
Kidd [Jk^a and Jk^b];
Kell [K1] and Cellano [K2];
Glycophorin B system S and s

There may be other systems of particular pertinence in different populations with a unique distribution of blood group antigens.

2.3. *Instrument standardization*

Details of the flow cytometer settings are not given since every instrument differs in the gains and voltage settings needed. Before use, the flow cytometer should be standardized against the appropriate fluorescent beads, available through the manufacturer. The forward and side scatter voltages for analysis of RBC are both on a linear scale, with fluorescence expressed on a logarithmic scale. Generally 50,000 events are counted, and because most analyses are single colour, compensation is generally not required.

2.4. *Standard protocol for quantitation of RBC surface antigens*

Five µl of well-mixed whole blood is washed with 2 ml of PBS and the supernatant removed to leave a button of red cells. When using a directly conjugated antibody, the appropriate volume of antibody is added according to the instructions from the manufacturer and the cells incubated in the dark for the suggested time. Controls should include at least samples of blood known to be positive and negative for the antigen of interest, and may need to include samples of heterozygous and homozygous antigen expression. For the very critical detection and quantitation of Rh(D) positive fetal cells in a Rh(D) negative maternal sample, a standard curve involving a series of samples of Rh(D) negative blood deliberately spiked with Rh(D) positive cord blood and covering a range between 0.1–4.0% heterogeneity should ideally be included in every assay. This type of standard curve is unlikely to be required for other antigen assays.

For unconjugated antibodies, 50 µl of the required antibody is added, briefly centrifuged (10 sec) to mix, and vortexed. After 30 min at room temperature (RT) the cells are washed once with 2 ml PBS and incubated for a further 30 min at RT in the dark with 50 µl of the secondary antibody, normally diluted 1:50 in PBS (dilutions will have to be titrated for reagents from any particular manufacturer). Positive and negative controls for the antigen of interest should be included, and as an additional negative control, a tube containing cells treated with the secondary antibody alone should be included. The samples are washed twice with 2 ml of PBS before resuspension in 2 ml PBS. The labelled cells are stable for at least 24 h if kept at 4° in the dark and should be briskly vortexed immediately before analysis.

2.5. *Standard protocol for quantitation of intramembrane antigen expression*

A relatively new flow cytometric test has been described [12] that can characterize RBC from patients with hereditary spherocytosis. The test does not involve an antigen/antibody reaction but depends on the relative fluorescent intensity of RBC stained with eosin-5-maleimide, a dye which reacts covalently with Lys-430 on the first extracellular loop of band 3 protein. Because of the presence of protein defects in the membrane of the red cells of patients with hereditary spherocytosis (HS), fluorescence of the affected RBC is significantly lower that that of normal cells, with a sensitivity of 92.7% and a specificity of 99.1%.

Eosin-5-maleimide (Molecular Probes Inc.) is dissolved in PBS at a concentration of 0.5 mg/ml and the stock solution stored in small aliquots at –20 °C, protected from the light. An aliquot is thawed immediately before the test, and 10 µl of fresh whole EDTA blood is incubated with 25 µl of the dye for 1 h in the dark. At least 6 normal samples must be included in every test run. The stained cells are washed 3 times in 2 ml of PBS and resuspended in 2 ml PBS before analysis. Standard beads should be used to calibrate the flow cytometer for the erythrocyte settings immediately before the analysis. The fluorescence intensity in mean channel fluorescence (MCF) units are determined for 50,000 events in the FL-1 (green fluorescence) channel for all samples.

3. Materials and methods, intracellular antigens

The most commonly used test dependent on an intracellular RBC antigen is the detection of high levels of haemoglobin F (HbF). Such a test has application in studies of patients suffering various thalassaemias and in the determination of possible fetomaternal haemorrhage when the mother is Rh(D) positive or of unknown blood group. It thus can replace the Kleihauer-Betke test [13], and has been shown to be a more reliable test [8, 9, 24], especially when the haemorrhage is large.

3.1. *Antibodies*

A single human monoclonal IgG antibody to HbF, directly conjugated, is available from at least two sources, Caltag Laboratories and Chemicon Australia. The antibody has been labelled with fluorescein by each individual manufacturer. An unconjugated antibody has been used in an indirect test by [24] [sheep anti-human hemoglobin F, IgG fraction, Calbiochem-Novabiochem Corp. followed by incubation with a secondary fluorescein-labelled donkey anti-sheep immunoglobulin (Chemicon Australia)]. In either the direct or indirect test, the red cell membrane must be fixed and permeabilized to allow access of the antibody to the interior of the cell. The antibodies above are stable to fixation by glutaraldehyde, but others, notably an antibody to HbF produced by Wallac Inc. must be used in conjunction with formaldehyde fixation.

3.2. *Fixation and permeabilization*

Glutaraldehyde is used as the fixative. The stock reagent is 25% (Sigma-Aldrich) which is freshly diluted 1:250 in cold PBS. One ml of diluted reagent is added to 10 µl of whole blood, the suspension vortexed thoroughly and allowed to stand for 10 mins at RT. The suspension is centrifuged and washed 3 times with 2 ml PBS. To the final button of fixed cells, 500 µl of 0.1% Triton X-100 (Sigma-Aldrich), diluted in PBS, is added and vortexed and allowed to stand for 3 min at RT. The cells are washed once in PBS and resuspended in 500 µl of PBS.

3.3. *Standard protocol for intracellular antigen expression*

An aliquot of 25 µl of the fixed and permeabilised RBC suspension is incubated for approximately 30 min with 10 µl of the directly conjugated antibody in the dark at RT. The cells are washed twice with 2 ml PBS before resuspension in approximately 500 µl PBS for analysis. Because the fixed RBC have somewhat different scattering characteristics compared to fresh cells, a specific programme on the flow cytometer should be dedicated for these tests. As with the tests for Rh(D) positive fetal cells against a background of Rh(D) negative maternal cells, at least one deliberately spiked sample should be included with each test run, and the gatings set using these positive controls. It must be emphasized that many maternal blood samples, and indeed in certain ethnic groups, large numbers of cells with high levels of HbF ('F'-cells) may be present and can lead to confusion. These cells rarely have the intensity of staining of true fetal cells however [24].

The availability of a PE-labelled anti-D and a FITC-labelled anti-HbF means that it is possible to combine the tests for fetomaternal haemorrhage. The Rh(D) antigen is rather labile in the presence of glutaraldehyde but destruction of the antigen can be avoided by incubating the cells with the anti-D first, then fixing and permeabilizing and finally incubating the cells with the anti-HbF [24]. The presence of the anti-D antibody already attached to the antigen protects it from destruction by glutaraldehyde.

3.4. *Routine determination of reticulocyte numbers*

Staining of reticulocytes with thiazole orange has become a routine quantitative flow cytometric test and perhaps hardly deserves a mention in this article. Commercial reagents are readily available, for example Retic-COUNT (Becton Dickinson Immunocytometry Systems). The dye thiazole orange stains fragments of DNA and RNA remaining in the reticulocytes and can be coupled with tests utilizing antibody to surface antigens, so that newly emerging reticulocytes derived from a donor graft may be distinguished from residual reticulocytes contained in transfused RBC or perhaps host reticulocytes. Thiazole orange staining is analysed in the green fluorescent channel, so blood group antigen staining must use either an anti-human immunoglobulin with a PE label for example, or a biotin label to be used in conjunction with a PE-streptavidin in a subsequent step.

4. Interpretation

4.1. *Rare event analysis*

Probably the most important clinical application of rare event analysis would be the detection and quantitation of fetomaternal haemorrhage, particularly in the case of Rh(D) negative mothers with an Rh(D) positive fetus. The presence of 0.25% Rh(D) positive fetal RBC in an Rh(D) negative maternal blood sample implies a haemorrhage of approximately 6 ml of fetal RBC across the placenta, or 12 ml of whole blood. Despite the somewhat loose assumptions of maternal blood volume and haematocrit inherent in the derivation of this formula for fetal RBC volume, it has been used effectively to calculate the volume of anti-D immunoglobulin that should be administered to the mother to avoid allosensitization. Not only has the flow cytometric test been of benefit in identifying women at risk of developing an anti-Rh(D) antibody, but it has had an additional benefit in helping to conserve prophylactic anti-D immunoglobulin, which is in short supply in many countries. Detection of fetomaternal haemorrhage in mothers who are Rh(D) positive does not have the same urgency, although it can confirm the clinical picture in cases of unexplained infant anaemia or stillbirth.

4.2. *Transplantation or transfusion chimaeras*

The monitoring of engraftment of erythrocyte progenitors following bone marrow or peripheral blood stem cell (PBSC) transplantation depends upon antigenic differences between the RBC of host and donor. Figure 1 illustrates the RBC antigenic changes in serial blood samples from a patient who received a mini-allograft from an HLA-matched sibling donor. The donor differed from the recipient in being Rh(D) negative and Jk^a positive. The initial rapid appearance of this latter antigen was due to the coincidental transfusion of two units of blood which were also Jk^a positive. The change in blood group of the patient took place over some 120 days but was finally complete. In contrast, Figure 2 illustrates the antigenic changes in the blood of a patient who was originally Rh(D) negative and received a mini-allograft from a sister who was Rh(D) positive. The patient presented with chronic graft-versus-host-disease 3 months after the transplant, and during treatment it became apparent that a fairly stable chimaeric state had been established with approximately half the RBC being derived from his donor. Flow cytometric analysis of blood samples from transplanted patients can thus confirm engraftment and may also give early warning of failure of the graft.

Figure 3 is a histogram of a flow cytometric analysis of a blood sample from a patient transfused with one unit of blood. This unit differed from his own blood group in being positive for the antigen s (an antigen within glycophorin B). It is on the basis of the antigenic make-up of the *majority* of red cells that decisions can be made as to the recipient blood group in patients who have been transfused before a full phenotype has been established.

In a further application of this type of analysis, minor antigenic differences in blood used for intrauterine transfusion or neonatal transfusion can serve as markers determining such parameters as the half-life of transfused cells, the level of fetal haemopoiesis and thus the timing of, or necessity for, further transfusions [25].

4.3. *Loss of blood group antigen expression and disease*

Paroxysmal nocturnal haemoglobinuria (PNH) is a disorder of the pluripotent stem cells resulting in a deficient expression of membrane-bound GPI-anchored proteins in different cell types. The most common of the GPI-linked red cell antigens to be routinely analysed by flow cytometry are CD55 (decay accelerating factor), CD59 (membrane inhibitor of reactive lysis) and to a lesser extent CD58 (LFA-3). A deficiency or absence of CD55 and particularly of CD59 renders the cells hypersensitive to complement lysis, and a wide variety of precipitating events such as infections, stress, transfusion or surgery can activate complement and

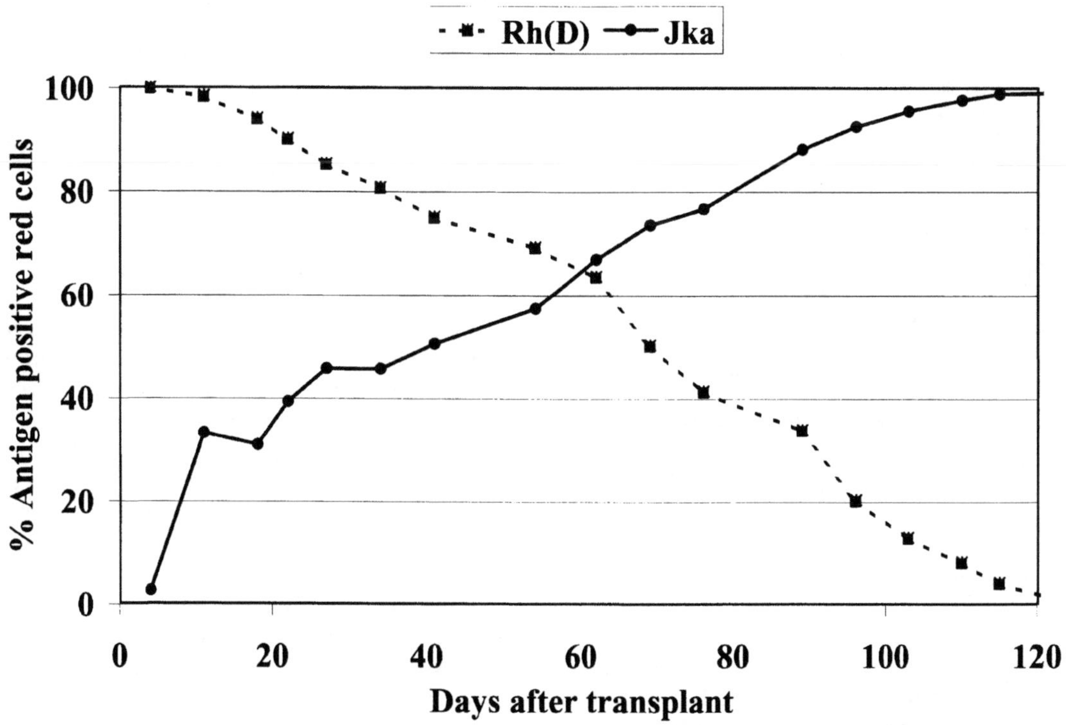

Figure 1. The changing pattern of red cell antigen expression during patient engraftment with peripheral blood stem cells from an Rh(D) negative, Jk^a-positive donor. The change in blood group was complete about 120 days post-transplant.

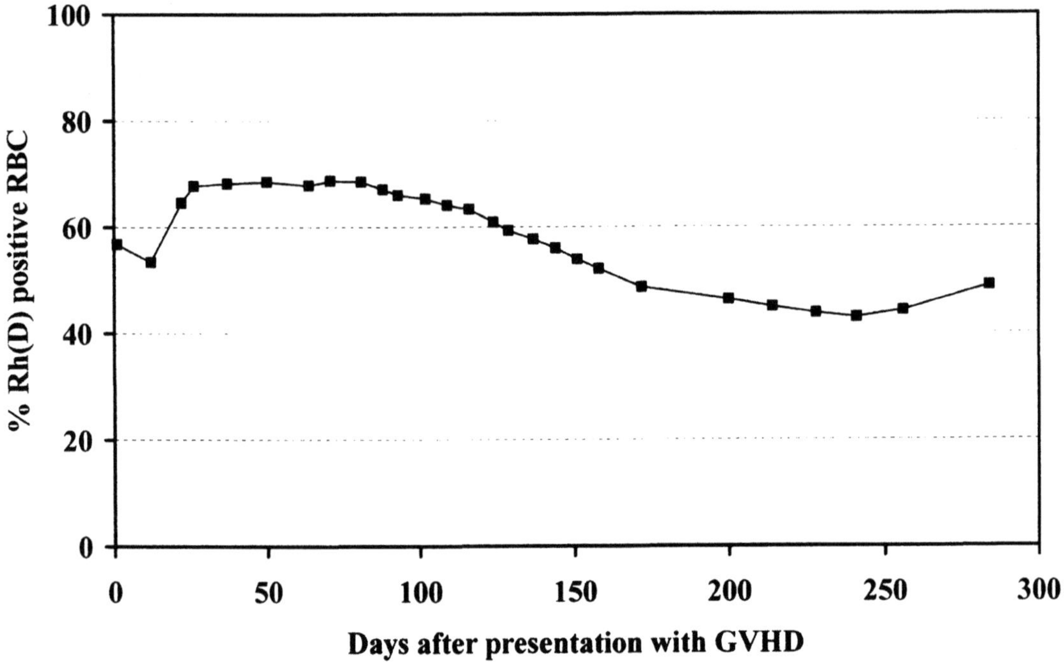

Figure 2. An illustration of a chimaeric state established in an Rh(D) negative patient who received a graft from an Rh(D) positive donor. Treatment for chronic graft-versus-host disease was begun approximately 80 days after the transplant, but the dual populations of red cells from host and donor have remained reasonably stable.

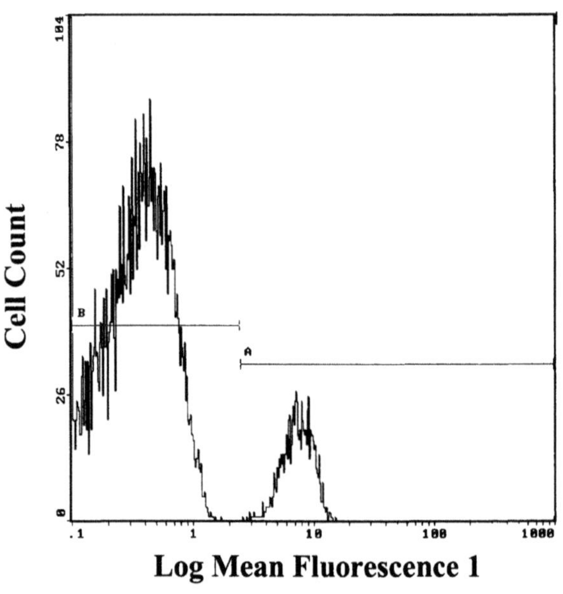

Figure 3. A flow cytometric histogram illustrating the detection of approximately 10% of RBC from a transfused single unit of blood unmatched for the blood group antigen s. The bulk of the cells are negative for this antigen, and thus represent the host RBC.

trigger intravascular haemolysis [19]. Expression of CD59 is high on normal RBC (Type I), and a clone of cells with reduced expression (Type II) or complete absence of CD59 (Type III) is very easily identified and may be said to define a diagnosis of PNH. In such a patient, all three types of cell may be present in any combination and in any proportion.

As mentioned in the Introduction, a progressive loss of expression of certain blood group antigens such as A, B, H and even Rh(D) occurs rarely in some malignancies such as acute myeloid leukaemia. Unlike the deficient cells in PNH, this loss of antigen expression does not in itself appear to cause problems, but does provide a serendipitous marker for monitoring the disease state and the success or otherwise of treatment.

5. Concluding remarks

The capability of accurately measuring mixed red cell populations has found many applications in transfusion and transplantation medicine. Antibodies to blood group antigens have provided a means to identify many of the individual populations, but the development of more antibodies to intracellular antigens like haemoglobin F should allow a widening of the range of differences that can be detected. Antibodies to intramembrane glycoproteins or glycolipids may be produced in the future that can distinguish foreign or aged red cells in the circulation. Already a dye has been discovered which can covalently stain intramembrane proteins and identify the defective red cells of patients with hereditary spherocytosis.

The technique of flow cytometry lends itself well to the detection and quantitation of red cell populations. It is sensitive, accurate and fast, the sample size need not be large, and interpretation of the results is straightforward.

References

1. Arnold R, Schmeiser T, Heit W, Frickhofen N, Pabst G, Heimpel H, Kubanek B (1986). Hemopoietic reconstitution after bone marrow transplantation. Exp Hematol 14: 271–277.
2. Austin EB, McIntosh Y, Hodson C, Lee D (1995). Anti-D quantification by flow cytometry: an alternative to the AutoAnalyser? Transfus Med 5: 203–208.
3. Banerjee D (1995). Flow cytometry related to hematopoietic malignancy. Transfus Sci 16: 315–320.
4. Blanchard D, Bruneau V, Bernard D, Germond-Arnoult F, Gourbil A, David B, Muller JY (1995). Flow cytometry analysis of dual red blood cell populations after bone marrow transplantation. Br J Haematol 89: 741–747.
5. Castilho L, Rios M, Bianco C, Pellegrino J Jr, Alberto FL, Saad ST, Costa FF (2002). DNA-based typing of blood groups for the management of multiply-transfused sickle cell disease patients. Transfusion 42: 232–238.
6. Cherif-Zahar B, Bony V, Steffensen R, Gane P, Raynal V, Goosens D, Laursen JS, Varming K, Jersild C, Cartron JP (1998). Shift from Rh-positive to Rg-negative phenotype caused by a somatic mutation within the RHD gene in a patient with chronic myelocytic leukaemia. Br J Haematol 102: 1263–1270.
7. Christensson M, Bremme K, Shanwell A, Westgren M, Christensson B (1996). Flow cytometric quantitation of serum anti-D in pregnancy. Transfusion 36: 500–505.
8. Davis BH, Olsen S, Bigelow NC, Chen JC (1998). Detection of fetal red cells in fetomaternal hemorrhage using a fetal hemoglobin monoclonal antibody by flow cytometry. Transfusion 38: 749–756.
9. Duguid JKM, Bromilow IM, Eggington J, Martlew VJ, McFadyen IR, Clarke CA (1996). Kleihauer testing and flow cytometry. A comparative study for assessment of feto-maternal haemorrhage. Hematology 1: 79–83.
10. Garretty G, Arndt P (1995). Applications of flow cytofluorometry to transfusion science. Transfusion 35: 157–178.
11. Jennings CD, Foon KA (1997). Recent advances in flow cytometry: application to the diagnosis of hematologic malignancy. Blood 90: 2863–2892.
12. King M-J, Behrens J, Rogers C, Greenwood D, Chambers K (2000). Rapid flow cytometric test for the diagnosis of membrane cytoskeleton-associated haemolytic anaemia. Br J Haematol 111: 924–933.
13. Kleihauer E, Braun H, Betke K (1957). Demonstration of fetal hemoglobin in erythrocytes by elution. Klin Wochenschrift 35: 637–638.
14. Kuhns WJ, Oliver RTD, Watkins WM, Greenwell P (1980). Leukemia-induced alterations of serum glycosyltransferase enzymes. Cancer Res 40: 268–275.
15. Kumpel B (2001). Labelling D+ RBCs for flow cytometric quantification of fetomaternal hemorrhage after the RBCs have been coated with anti-D. Transfusion 41: 1059–1063.
16. Lazarus HM, Chahine A, Lacerna K, Wamble A, Iaffaldano C, Straight M, Rabinovitch A, Schimenti KJ, Jacobberger J (1992). Kinetics of erythrogenesis after bone marrow transplantation. Am J Clin Pathol 97: 574–583.
17. Lynen R, Neuhaus R, Schwarz DW, Simson G, Riggert J, Mayr WR, Kohler M (1995). Flow cytometric analyses of the subclasses of red cell IgG antibodies. Vox Sang 69: 126–130.
18. Michelson AD (1996). Flow cytometry: a clinical test of platelet function. Blood 87: 4925–4936.
19. Min H, Thirumala SD, Reid ME (1999). Laboratory diagnosis of paroxysmal nocturnal hemoglobinuria: a review. Immunohematology 15: 144–149.
20. Nance SJ, Nelson JM, Arndt PA, Lam HC, Garratty G (1989). Quantitation of fetal-maternal hemorrhage by flow cytometry. A simple and accurate method. Am J Clin Pathol 91: 288–292.
21. Navenot JM, Bernard D, Harousseau JL, Muller JY, Blanchard D. (1996). Expression of glycosyl-phosphatidylinositol-linked glycoproteins in blood cells from paroxysmal nocturnal haemoglobinuria patients. A flow cytometry study using CD55, CD58 and CD59 monoclonal antibodies. Leuk Lymph 21: 143–151.
22. Nelson M, Popp H, Horky K, Forsyth C, Gibson J (1994a). Development of a flow cytometric test for the detection of D-positive fetal cells after fetomaternal hemorrhage and a survey of the prevalence in D-negative women. Immunohematology 10: 55–59.
23. Nelson M, Forsyth C, Popp H, Gibson J (1994b). Rapid detection of Rh(D)- or K-positive fetal red cells in chorion villus samples by a flow cytometric technique. Transfus Med 4: 297–302.
24. Nelson M, Zarkos K, Popp H, Gibson J (1998). A flow cytometric equivalent of the Kleihauer test. Vox Sang 75: 234–241.
25. Nelson M (1999). An overview of the use of flow cytometry in the analysis of mixed red cell populations. Pathology 31: 191–198.
26. Packman CH (1998). Pathogenesis and management of paroxysmal nocturnal hemoglobinuria. Blood Rev 12: 1–11.
27. Pirofsky B, Rosner ER (1974). DTT test: a new method to differentiate IgM and IgG erythrocyte antibodies. Vox Sang 74: 480–488.
28. Popp HJ, Nelson M, Forsyth C, Falson M, Kronenberg H (1995). Quantifying the loss of ABO antigenicity in a patient with acute myeloid leukemia by flow cytometric analysis. Immunohematology 11: 5–7.
29. Tomer A (1997). A sensitive and specific functional flow cytometric assay for the diagnosis of heparin-induced thrombocytopenia. Brit J Haematol 98: 648–656.
30. Van der Meulen FW, de Bruin HG, Goosen PC, Bruynes EC, Joustra-Maas CJ, Telka PHG, von dem Borne AE, Engelfreit CP (1980). Quantitative aspects of the destruction of red cells sensitized with IgG_1 autoantibodies: an application of flow cytofluorometry. Brit J Haematol 46: 47–56.
31. Wright JF, Lazarus AH, Freedman J (1995). Applications of flow cytometry in the analysis of blood leukocytes. Transfus Sci 16: 333–341.

Methods in Cell Science 24: 27–35 (2002).

Chromosome sorting and genomics

L. Scott Cram, Carolyn S. Bell & John J. Fawcett
National Flow Cytometry Resource, Los Alamos National Laboratory, Los Alamos, New Mexico, 87545, USA

Flow Cytogenetics describes the application of flow cytometry and sorting to the analysis of mammalian chromosome classification and purification, respectively. Early publications on flow cytogenetics appeared in 1975 and described single color (univariate) analysis of human and Chinese hamster chromosomes. Since that time enormous progress has been made in the analysis and sorting of chromosomes using two color (bivariate) analysis methods. Applications include high speed sorting for construction of chromosome specific libraries that established feasibility for the human genome program and the construction of chromosome specific fluorescent in-situ hybridization (FISH) probes for research and clinical applications. Additional applications include: quantification of radiation induced chromosome damage, determination of chromosome aneuploidy for somatic cell genetic studies, and detection of numerical and structural aberrations associated with genetic diseases such as Down's syndrome (Table 1).

Flow karyotyping requires isolation of intact metaphase chromosomes, staining the chromosome suspension with a fluorescent tag, and rapid quantitative analysis in a flow cytometer (Figure 1). Chromosome isolation consists of freeing individual chromosomes from mitotic cells and stabilizing their structure. Staining reactions are designed to label a mixture of chromosome types so that one chromosome type is distinguished from another. The critical steps of how to achieve chromosome isolation, stabilization, staining and analysis are contained in the appendix to this chapter. The ultimate goal is to resolve each chromosome type from any given species. In the case of chromosomes isolated from human cells, this means resolving 23 populations when using cells of female origin (22 auotsomes, XX) and 24 populations in cells of male origin (22, XY).

Table 1. Applications of Chromosome Analysis and Sorting

Analysis
Monitor progressive karyotype changes
Quantitate chromosome polymorphisms
Detect variant 'marker chromosomes'
Dosimetry
Cytogenetic screening
Classification of chromosomes from different species
Sorting
Construction of chromosome specific libraries
– gene Mapping
– linkage and pedigree analysis
– study interspecies gene homology and evolution
– analysis of chromosome rearrangements
– chromosome-specific labels for flow karyotyping
Spot blot sorting - gene mapping
Identification of peaks in flow karyotype
Reverse chromosome painting

Several independent variables can be measured for each chromosome as it passes through the interrogation (analysis) region of a flow cytometer at typical analysis rates of 1,000–5,000 chromosomes/second. Flow karyotype analysis complements classical cytogenetics in many ways and the two techniques are often used in parallel. Flow karyotyping provides quantitative, precise information about chromosome properties, such as DNA content for several hundred thousand chromosomes. The attractive advantages of flow karyotype analysis are based on its quantitation, statistical precision and flexibility. Other chromosome properties, such as chromosome length, RNA content, total protein and specific protein content have also been measured (Table 2).

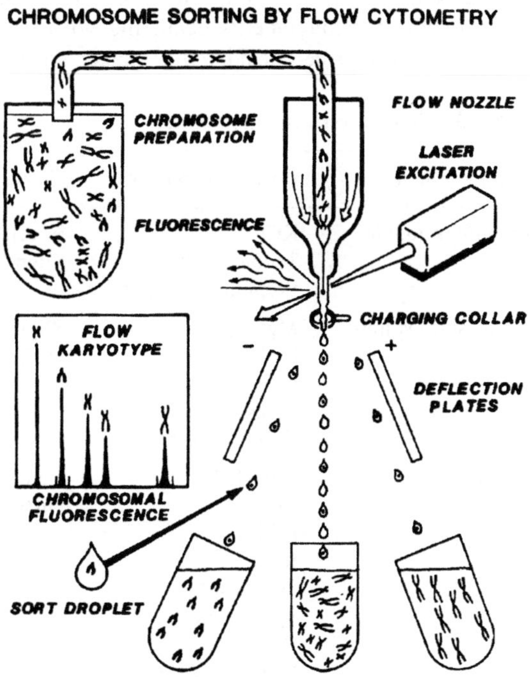

Figure 1. Schematic of chromosome analysis and sorting: flow cytogenetics.

Table 2. Flow karyotype analysis

Description	Probe(s)
Length	Pulse width
Centromeric index	Pulse shape
Elasticity	Pulse width
Proximity relationships	Energy transfer
DNA Content	PI, EB, Hoechst, DAPI, Acridine Orange
Base composition	Hoechst 33258, CA3, Distamycin A, Netropsin,
RNA	Pyronin Y, Acridine Orange
Proteins	SR 101, FITC, O-phthaldehyde
Histones	Antibodies
Chromosome Specific	Monoclonal Ab, FISH probes
Replication	BrdU and Hoechst
Molecular Fluidity	Hoechst 33342
Polymorphisms	CA3 and Hoechst, Distamycin and CA3

The ability to resolve all chromosome types from any mammalian species usually depends upon differences in inter-chromosomal DNA content (either total DNA content or base pair ratios) and instrumental resolution. These measurements will be discussed in terms of how they impact the application of the technology. Applications of univariate (one color) flow karyotype analysis include: determining and monitoring karyotype instability, variation in the frequency of a chromosome type (due to the occurrence of trisomy or monosomy), chromosomal polymorphisms that can be resolved and quantitated where it is not other wise possible, and chromosome rearrangements occurring with sufficient frequency to be considered a 'maker chromosome'. This is particularly important in a heteroploid population where a very large variety of chromosome types are observed. Chromosome sorting is used to identify chromosome types in a flow karyotype and has been extensively used for gene mapping, cloning, and molecular characterization of normal and rearranged chromosomes.

1. Univariate techniques for high-resolution flow karyotype analysis

Protocols for chromosome isolation, staining and analysis have been developed to the point where univariate flow karyotypes for several species allow all chromosome types to be resolved as separate peaks. Critically important is the selection of appropriate buffers for stabilizing and staining chromosomes. Table 3 illustrates three different buffers and the resulting characteristics of each.

The choice of buffers depends of the eventual application of the procedure. Generally, the propidium iodide procedure is used for one-color staining procedures resulting in excellent morphology and high-resolution measurements. Magnesium sulfate is ideal for small samples and two color staining, it's only draw back being reduced molecular weight. The polyamine procedure is the method of choice for two-color staining of human chromosomes for sorting applications. While the chromosomes do not maintain their normal morphology this is not a problem for most applications.

High-resolution univariate flow karyotypes are produced using the specialized procedures presented in the appendix. Figure 2 illustrates the excellent resolution that can be achieved using the propidium iodide staining procedure. Each chromosome type of

Table 3. Chromosome isolation buffers

Common name	Buffer components	Chromosome morphology	Protocol choice for
Propidium iodide	KCl Triton X-100 Propidium iodide*	Excellent morphology Slightly extended	High resolution One color analysis Fast and easy to reproduce
Magnesium sulfate	$MgSO_4$ HEPES Dithiothreitol RNase Triton X-100	Slightly contracted Difficult to recognize	High resolution, Flexible for different staining protocols, Small sample size Preserves chromosomal proteins
Polyamine	Tris-HCl EDTA EGTA KCl Spermine* Spermidine* Digitonin Mercaptoethanol	Contracted Difficult to recognize	Cloning procedures, High resolution, Two color staining, Maintains high molecular weight

* Designates purported stabilizing agent.

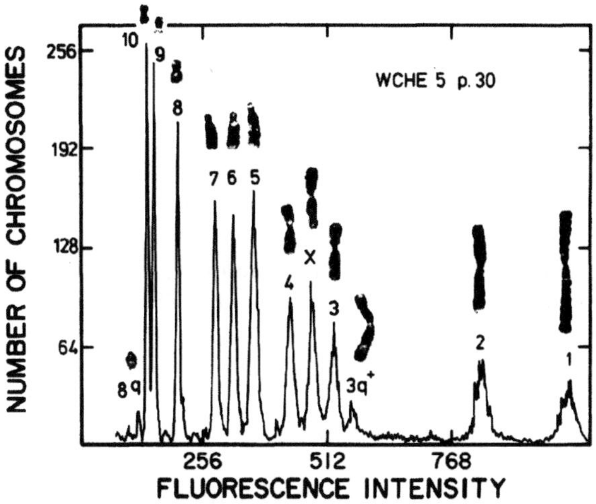

Figure 2. Whole Chinese hamster chromosomes stained with propidium iodide.

this Chinese hamster cell strain can be resolved as well as a chromosome (number 3) with an insertion element in the long arm, and a marker chromosome (8q). The precision of this measurement can be summarized as follows.

An increase or decrease in chromosome number can be detected when as few as 10% of the cells contain an extra chromosome type (partial trisomy). Shifts in DNA content (peak position) are detected with a few percent change in fluorescence intensity. Uniformity of chromosome illumination during flow analysis is critical for high-resolution measurements. This can be accomplished by keeping the sample concentration as high as possible so that the chromosome analysis rate can be kept high while maintaining a small (< 2 μm) sample stream diameter and spreading the laser beam so that variation in intensity across the sample stream is less than 1%. A sample concentration of about 1–5 × 10^6 chromosomes/ml is maintained. Figure 3 illustrates analogous data for cloned Chinese hamster embryo cells where resolution of the peaks resolves the two number 9 chromosomes based on a slight difference in their DNA content. The coefficient of variation of the peaks is on the order of 1–1.3%.

The type of amplifier used will influence the appearance of a distribution. Figure 4 compares the visual differences that result from the use of a linear or logarithmic amplifier. This data set was acquired by splitting the signal from the preamplifier and sending the two identical signals to linear (top figure) and logarithmic amplifiers (bottom figure). The information in both distributions is exactly the same. Data acquisition with a logarithmic amplifier offers advantages for visual interpretation. Chromosomes from a truly euploid cell (no aberrations and two copies of each homologue) will, when analyzed using a logarithmic amplifier, have all peaks of equal height and equal width. Chromosomes present in only one copy (chromosome X) and homologues with a difference in DNA content (chromosome 9) are readily detected as their peak heights are about half as high as chromosome types present in two copies per cell.

2. Bivariate flow karyotyping

For univariate flow karyotyping, chromosome discrimination is based on the amount of fluorescent dye bound to the chromosome. Many of the fluorochromes used for flow karyotyping bind only to

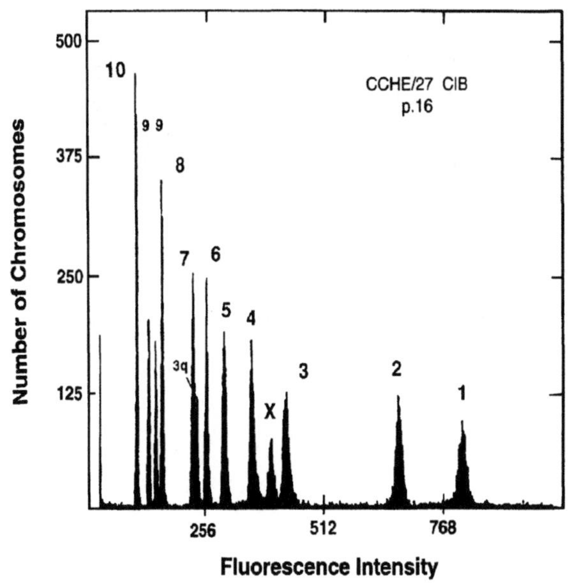

Figure 3. Cloned Chinese hamster embryo chromosomes.

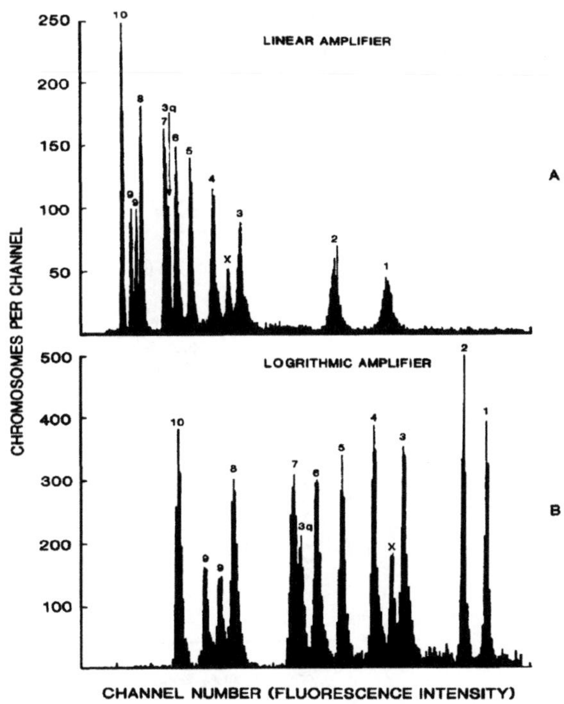

Figure 4. Comparison of linear and log amplification.

nucleic acids so that discrimination is largely based on total DNA content. This classification procedure is limited when the number of chromosome types exceeds what is found in species such as Chinese hamster. Bivariate flow karyotyping (chromosome classification based on two fluorochromes) was developed to take advantage of the fact that some dyes like Hoechst 33258 (HO) and Chromomycin CA3 (CA3) bind preferentially to adenine- thymine (AT) or guanine-cytosine (GC) rich DNA, respectively. This pair of fluorochromes allows classification of chromosomes according to DNA content and DNA base composition. HO can be excited by the ultraviolet (UV) (351 and 353 nm) beam from an argon ion laser. It fluoresces around 450 nm. CA3 can be excited by the 458 nm beam from an argon ion laser. It fluoresces around 520 nm. Substantial energy transfer occurs from HO to CA3 so that most fluorescence occurs around 520 nm. Neither dye binds covalently so that DNA can be isolated from chromosomes stained with these dyes; the dyes can be removed and the DNA processed using conventional molecular biological techniques.

Once stained, chromosomes are measured using dual-beam flow cytometry. The chromosomes flow sequentially through laser beams adjusted to the UV to excite HO and 458 nm to excite CA3. Fluorescence resulting form the excitation is recorded as a measure of the HO dye content and fluorescence from the 458nm excitation is recorded as a measure of the CA3 content. Thus a correlated pair of measurements are made for each chromosome. Measurements are typically made from 10^4 to 10^6 chromosomes. Figure 5 illustrates a flow karyotype for a normal male displayed as a contour plot. The numbers identify the chromosome contained within the peak. Flow karyotypes for normal human cells typically show one peak for each chromosome type except chromosomes 9–12 and sometimes 14 and 15. There is a continuum of events that underlies the various peaks in a bivariate flow karyotype that is caused by chromosome and nuclear debris fragments and chromosome clumps that are produced during the chromosome isolation process. Typically, background debris is most prevalent underneath the peaks of lowest fluorescent intensity and extends as a band of decreasing magnitude along the diagonal of the flow karyotype.

3. Chromosome sorting

Chromosome purification by sorting has facilitated both gene mapping and production of chromosome-specific recombinant DNA libraries. These applications require the highest possible discrimination of chromosome types one from the other and from chromosomal debris and clumps. If small numbers of sorted chromosomes are needed, chromosomes can be processed at sufficiently low rates to maintain very high measurement resolution and high sorting purity. In such situations, chromosome purity of greater than95% can be obtained from normal human cells. Many applications require millions of sorted chromosomes, for these applications, some resolution must be sacrificed for sorting speed. Hamster- human hybrid cell lines have proven particularly useful as sources of human chromosomes. Hybrid cell lines are critical for sorting human chromosomes 9–12 and 14–15 due to their overlap in DNA content (Figure 5). However chromosome polymorphisms occurring in some human cells allow sorting of chromosomes such as chromosomes 14 and 15. A preferred option, is illustrated in Figure 6, the use of somatic cell

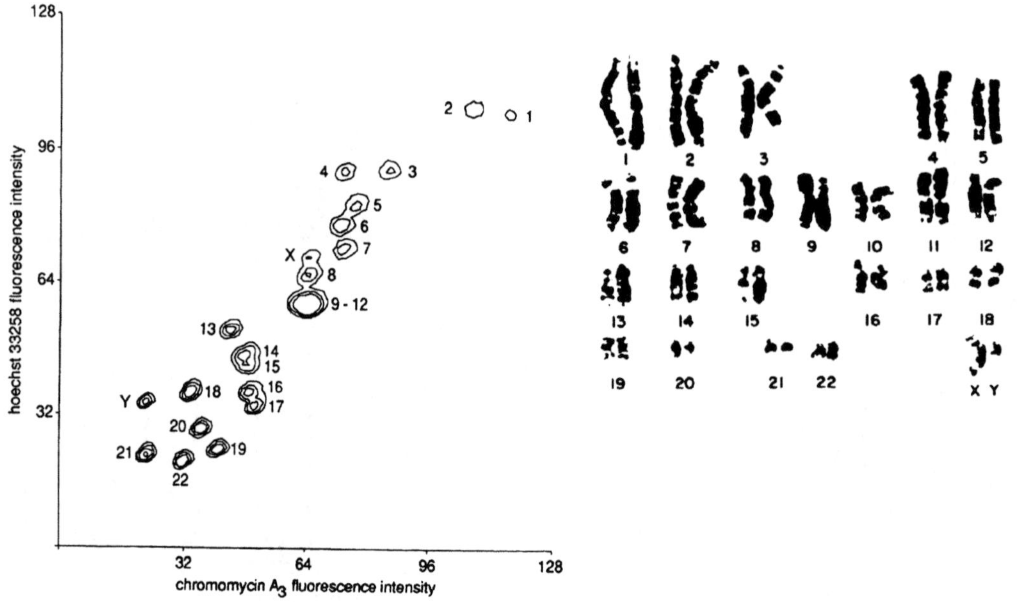

Figure 5. Bivariate flow karyotype of normal human cells and the corresponding Giemsa banded karyotype.

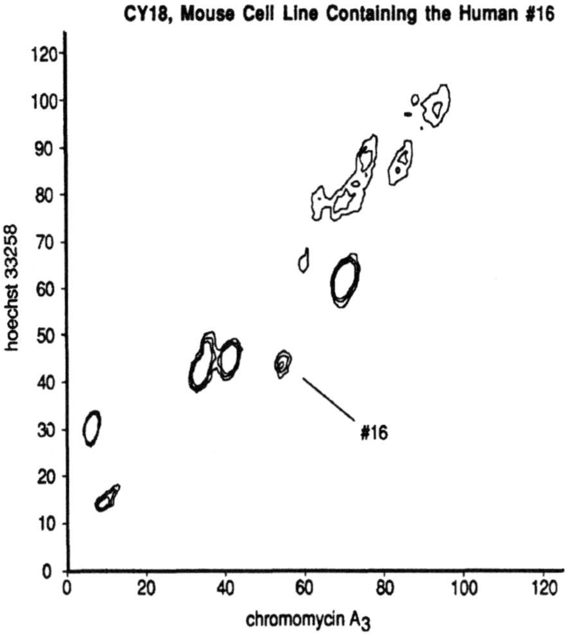

Figure 6. Bivariate flow karyotype of a mouse/human somatic cell hybrid.

hybrids. The advantage of using somatic cell hybrids for sorting a single human chromosome (for example, chromosome 16) from a mouse cell line is that only a single human chromosome is present and very high purity can often be obtained. However, hybrid cell lines often contain a partially deleted human chromosome.

The number of chromosomes typically sorted for different applications varies enormously. For example for polymerase chain reaction applications as few as 100–300 chromosomes are sorted. For phage cloning, 0.25 μgm of DNA; cosmid cloning, 0.5 μgm of DNA; yeast artificial chromosomes (YAC's) 1.0 μgm of DNA, and for bacterial artificial cloning (BAC's) > 1 μgm of DNA is required.

Depending on the purity required and the resolution of the chromosome type(s) being sorted, sorting times can require days or weeks of sorting time on a conventional speed sorter. A good figure of merit for high purity sorting on a conventional flow sorter is about 10^6 chromosomes in 6 hours. This is assuming standard sorting conditions that we define as an analysis rate of 1,250 chromosomes/second, a homolog flow rate of 52/second, a coincidence rate of 5/second, and a chromosome number concentration of 2.4×10^7/ml. High speed sorting of chromosomes can increase this rate by roughly a factor of 5 without seriously diminishing resolution or recovery. At faster analysis rates resolution is seriously degraded, as is purity.

4. Cytometric technologies for the human genome program

Chromosome sorting and analysis played a major role in the early stages of the human genome program. New genome related applications continue to evolve in the areas of genomics and proteomics. Five major areas of application have developed: flow cytogenetics, construction of chromosome specific libraries, bead based assays for detection of single nucleotide polymorphisms (SNP's), DNA fragment analysis, and single molecule DNA sequencing.

Flow cytogenetics. Clinical applications have evolved around the ability to detect and sort aberrant chromosomes. In particular the identification of translocations by the application of chromosome specific probes derived from sorted chromosomes. This is a particularly powerful technique that has

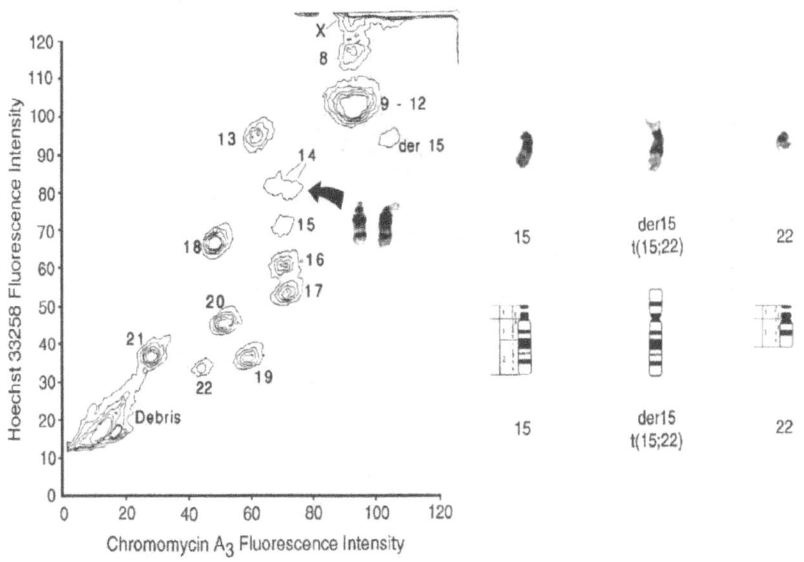

Figure 7. Flow karyotype of a patient with a translocation, t(15; 22).

seen limited application due to the commitment of resources necessary to make it a simple routine clinical technique. Figure 7 illustrates the type of application to which chromosome analysis and sorting can be used. In this example chromosomes larger than number 8 were placed off scale in order to spread out the lower half of the flow karyotype. Cytogenetic analysis confirmed a translocation of chromosome 15 onto chromosome 22, resulting in a new peak (der 15) located just below the 9–12 chromosome peak and leaving a single chromosome 15 and a single chromosome 22. In addition, the two number 14 chromosomes in this individual contain a significant difference in DNA content as illustrated by the nonsymmetrical shape of the number 14 contour.

Two approaches to understanding the flow karyotype illustrated in Figure 7 have been successfully used, both involve sorting the abnormal peak found in the bivariate flow karyotype. Giemsa banding of chromosomes prepared using the polyamine buffer is not realistic based on the highly condensed appearance of the chromosomes. The preferred technique is to sort the aberrant peak and to prepare a FISH probe from the sorted material and hybridize the sorted (and PCR amplified) DNA to a normal chromosome spread, or to prepare a spread of the sorted material and hybridize normal chromosome specific probes (FISH) to the sorted fraction. These two FISH techniques yield complimentary information useful to the cytogeneticist seeking to identify rearrangements in terms of the chromosome of origin and the chromosome of destination.

Construction of chromosome specific libraries. The single largest application of chromosome sorting has been the creation of chromosome specific libraries. Human chromosome specific libraries provided the initial starting material that was used in the early stages of the human genome project. Sorted human chromosomes, as well as other species, have been cloned using several vectors as diagramed in Figure 8. The availability of human libraries constructed from a signal human chromosome simplified the project by being able to assign and map DNA sequences known to have come from a single chromosome type.

Cosmid vectors can accept a 40 kb fragment of DNA. DNA of this size and molecular weight can be achieved in a flow cytometer by taking appropriate steps. Successful sorting is accomplished using the polyamine buffer as the sheath (see Appendix), accumulating a packed pellet of material, and maintaining the sorted material at wet ice temperatures.

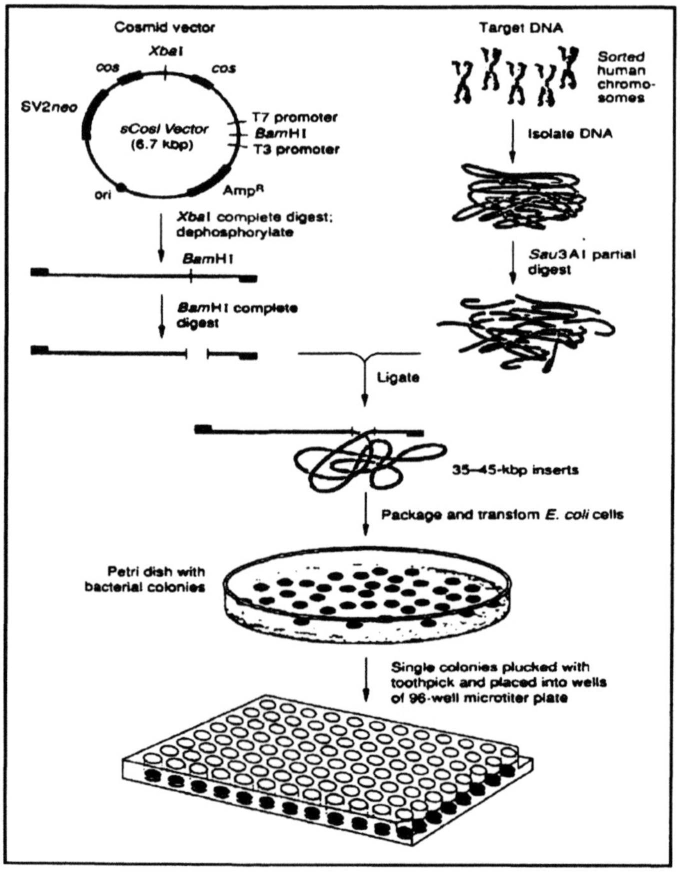

Figure 8. Cosmid cloning of sorted human chromosomes.

Because sorting times are long, machine maintenance is critical to avoid bubbles, debris, room temperature fluctuations or anything that might cause the analysis and sorting conditions to be altered.

Bead based assays, DNA fragment analysis, and single molecule DNA sequencing. All three of these relatively new developments further demonstrate the versatility of flow cytometry to measure and analyze genetic changes at the molecular level. Bead based flow cytometric assays are being used to detect single nucleotide polymorphisms (SNP's). DNA fragments in the size range of a few hundred bases have been analyzed in a specialized flow cytometers capable of photon counting. The photon counting flow cytometer is used in conjunction with very high quantum yield DNA fluorochromes and slow flow to analyze bacterial DNA that has been cut into fragments with well characterized restriction enzymes. In addition, all the necessary components of single molecule DNA sequencing have been demonstrated using specialized flow cytometers to rapidly sequence very long DNA segments (40 kb). The technique requires suspending a fluorescent stained strand of DNA from an optically trapped bead. When positioned in a flowing stream to which a processive enzyme has been added, sequential cleavage of single nucleotides occurs. Once cleaved, the labeled nucleotides flow through a laser beam and are analyzed based on their fluorescence signature. While this capability has not yet been reduced to practice all the essential elements are in place to be able to rapidly sequence very long DNA segments.

5. Summary

Flow cytogenetics is an exciting field that has made numerous contributions to our understanding of the human genome and karyotype instability. The results of these studies have been fruitful as they have been used to augment molecular, cytogenetic, and cell based investigations. The field continues to evolve into new and exciting areas as evidenced by the development of new applications for SNP detection and DNA fragment analysis. These assays will have an important role to play in understanding human heterogeneity and for the analysis of biological agents.

6. Further reading

- Flow Cytogenetics, (J. Gray, ed.), Academic Press, New York, 1989.
- J. W. Gray and L. S. Cram, Flow Karyotyping and Chromosome Sorting. In: Flow Cytometry and Sorting, 2nd Edition (M. Mendelsohn, T. Lindmo, and N, Melamed, ed.) Wiley-Liss Inc. New York, pp. 503–529 (1990).
- L. S. Cram, M. Campbell, J. Fawcett, and L. L. Deaven, Polyamine Buffer for Bivariate Human Flow Cytogenetic Analysis and Sorting. In: Methods in Cell Biology, Vol. 33, pp. 377–383 (1990).
- L. S. Cram, F. A. Ray, and M. F. Bartholdi, Univariate Analysis of Metaphase Chromosomes Using the Hypotonic Potassium Chloride-Propidium Iodide Protocol. In: Methods in Cell Biology, Vol. 33, pp. 369–377 (1990).

Acknowledgements

Collaborators at the National Flow Cytometry Resource, Los Alamos National Laboratory, who have made outstanding contributions to our development of flow cytogenetics include James Jett, Carolyn Bell, Joe Fawcett, John Martin, and Larry Deaven. This work was conducted under the auspices of the Department of Energy and with support from the National Institutes of Health, Division of Research Resources, Grant RR01315.

Appendices

Protocols for chromosome isolation and staining.

Appendix 1. Preparation and Analysis of Chinese Hamster Chromosomes with Propidium Iodide

Preparation of chromosomes typically involves the following steps:

1. Arrest cells in metaphase with colcemid.
2. Collect mitotic cells by shake off.
3. Swell the cells with hypotonic KCl.
4. Rupture the cell membrane with detergent and disperse chromosomes by forcing suspension through a syringe needle.
5. Stain the chromosome with DNA specific dye(s).
6. Analyze preparations.

Details:

1. Colcemid block a subconfluent and exponentially growing monolayer of cells. Concentration depends on cell type and varies between 0.01 μg/ml to 0.1 μg/ml. Generally one to two T-150 flasks that are 50–60% confluent yield a good preparation when using Chinese hamster cells.
2. Incubate cells in colcemid for 3 hours, 37 °C. Gently shake off mitotic cells and recover by centrifugation (100 ×*g* for 6 min).
3. Prior to centrifugation remove a small amount of media and stain with 10 mM Hoechst 33342 and determine total cell number and percentage mitotic cells.
4. Remove **ALL** tissue culture media, resuspend the pellet.
5. Add 0.5 ml of hypotonic KCl (75 mM) containing 50 μg/ml propidium iodide.

6. Allow cells to swell for 10 minutes at room temperature.
7. Monitor cell swelling in a microscope. Ideal conditions result in mostly intact cells with intact membranes and are therefore, non-fluorescent.
8. To 0.5 ml of cells, add 0.25 ml of the following: 75 mM KCl, 1% Triton X-100, 50 μg/ml PI, and 1 mg/ml Rnase.
9. Incubate 3 minutes at room temperature to partially dissolve cell membrane.
10. Syringe cell suspension 4-5 times through a 1″ 22 g needle with the point of the needle against the side of the tube.
11. Incubate chromosome suspension for 30 minutes at 37 °C and filter with 60 μm mesh.
12. Isolated chromosomes are ready for analysis, held at 4 °C they can be maintained for at least several days.

Appendix 2. MgSO4 Chromosome Isolation Procedure

1. Make up buffer immediately before use: for 10 ml:

1.0 ml	100 m*M* $MgSO_4$ (final concentration 10 m*M* at pH 8.0)
9.0 ml	55 m*M* KCl + 5.5 m*M* HEPES @ pH 8.0 (should be prepared no more than 1 week in advance) (final concentration 50 m*M* KCl, 5 mM HEPES)
0.5 ml	RNase stock: 3.0 mg/ml in 50 m*M* KCl of Worthington at 47330 units/mg (final conc 0.13 mg/ml)
0.25 ml	dithioerythriotol (120 m*M*) in water (prepared and aliquoted in small amounts and frozen until use) (final conc. 3 m*M*)

2. Filter above solution through 22 μm Millipore (disposable) filter. Arrest cells in metaphase with colcemid
3. Shake off mitotic cells. Distribute into tubes, 3 ml mitotic cells/tube is usually sufficient.
4. Spin at 4 °C, 8 min, 800 rpm.
5. Remove all supernatant.
6. Flick bottom of tube several times to loosen pellet.
7. Add 1 ml isolation buffer as prepared in step 1. Flick several times to mix.
8. Let stand for 10 min at room temperature.
9. Add 0.1 ml (to each 1 ml of sample) triton X-100 (stock solution 2.5% in H_2O; filtered through 2 μm disposable filter): final concentration 0.25%
10. Let stand for 10 min at room temperature.
11. Syringe 3× with a 22 gauge needle (1-1/2″ long) using 3 ml size for 1ml volume; 5 ml syringe for 3 ml sample.
12. Pool tubes if necessary.
13. Incubate 37° for 30 min in water bath.
14. Stain with DNA stain: use 20 μg/ml PI or 2 μg/ml Hoechst 33258 or 20 μg/ml CA3.

Appendix 3. Polyamine Buffer for Bivariate Human Flow Cytogenetic Analysis and Sorting

I. *Materials*

A. *Stock solutions:*

Stock solution I: Prepare a 10× solution (stock solution I) containing Tris, EDTA, KCl, and NaCl.

150 mM Tris-HCl	(Trizma Hydrochloride, Sigma cat. #T-3253)
20 mM EDTA	(Tetrasodium salt, Sigma cat. # ED455)
800 mM KCl	
200 mM NaCl	

Combine 2.36 gm Tris-HCl, 0.760 gm EDTA, 5.96 gm KCl, and 1.16 gm NaCl in 100 ml distilled water, filter sterilize, store at room temperature, and make fresh every month.

Stock solution II: Prepare a 10× solution (stock solution II) of EGTA

5 mM EGTA	(Sigma cat. #E-4378)

Add 0.190 gm to 100 ml distilled water, add as little concentrated NaOH as possible to get into solution, filter sterilize, store at room temperature, and make fresh every month.

Polyamines: Make up 10 ml and freeze in 0.1 ml aliquots, do not refreeze after use.

0.4 M Spermine tetrahydrochloride (Sigma cat # 5677), use 139 mg/ml in dH_2O
1.0 M Spermidine trihydrochloride (Sigma cat # 56766), use 254 mg/ml in dH_2O

Ohnuki's hypotonic swelling buffer:

55 mM Sodium nitrate	5 ml
55 mM Sodium acetate	2 ml
55 mM Potassium chloride	10 ml

Just before swelling the mitotic cells, add spermine and spermidine to the hypotonic swelling buffer in the same proportions as added to the chromosome isolation buffer (12.5 μl/25 ml)

B. *Chromosome isolation buffer (just prior to use):*

1. Add 5.0 ml of each of the two 10× stock solutions to 40 ml distilled water and adjust pH to 7.2 ± 0.05 with HCl. Add 50 μl of 2-Mercaptolethanol.
2. To 25 ml of the above solution, add 30 mg digitonin and incubate at 37 °C for 45 min.
3. Sterile filter (0.2 μm). Undissolved digitonin is removed in this step.
4. Add 12.5 μl of spermine and 12.5 μl of spermidine.
5. Place on ice.

II. *Isolation and Staining Procedure (human fibroblasts)*

1. Block cells with 0.1 μg colcemid/ml of tissue culture media for 12–15 hours.

2. Prepare fresh chromosome isolation buffer and place on ice.
3. Collect suspension cells in 50 ml centrifuge tubes, at 40 °C.
4. Adjust supernatant volume to achieve a cell concentration of 8×10^6 mitotic cells per tube.
5. Centrifuge at 100 ×*g* (approximately 800–1,000 rpm) for 8 min at 4 °C.
6. Aspirate supernatant, leaving as little as possible behind. Flick tube to loosen pellet.
7. Cell swelling.
   a. Add 5ml of Ohnuki's hypotonic swelling buffer per tube.
   b. Place cells into 15 ml polypropylene centrifuge tubes at a cell concentration of 8×10^6 cells/tube. Cells are swollen at room temperature for 70 minutes.
8. Centrifuge at 100 ×*g* for 5 minutes, room temperature.
9. Aspirate supernatant and flick tube to loosen pellet.
10. Add 1.0 ml of cold chromosome isolation buffer to each tube. Mix gently.
11. Vortex each tube vigorously for 30 sec. Place sample on ice for 10 min and observe chromosomes in microscope using propidium iodide to determine degree of dispersion. If needed, repeat vortexing for 10 sec and maintain sample on ice while rechecking dispersion. Repeat vortexing cycle as needed.
12. Chromosomes can be stained at this point or stored unstained at 4 °C.
13. Staining is with Hoechst 33258 and chromomycin A3. Hoechst stock solution is made up in distilled water (0.5 mg/ml) and 4.5 µl added to 1.0 ml of chromosomes (final concentration, 3.12 µM). Chromomycin A3 is made up in McIlvane's buffer pH 7, with 5 mM $MgCl_2$ (1:1). 150 µl is added to 1.0 ml of chromosomes at least three hours before analysis to establish equilibrium (final concentration, 55 µM).

Appendix 4. Polyamine Sheath Buffer – LARGE Volume Recipe

Solution A EGTA

- Place approx. 1.8 liters of water in a 2 liter flask.
- Add 3.8 g EGTA, Anhydrous Mol. Wt. 380.4, Sigma Cat No. E-4378 EGTA.
- Place flask on a magnetic stirrer and begin dissolving the EGTA.
- Add only enough NaOH to dissolve the EGTA. If the EGTA is not in solution after 15 minutes, add 1 pellet of NaOH every 5–10 minutes until the EGTA is dissolved.
- Add additional water to make 2.0 liters and mix well.
- Store in the refrigerator.

Solution B EDTA

- Place approx. 0.9 liters of water in a 1 liter flask.
  Add: 119.28 g KCI
  47.28 g Tris-HCl, Anhydrous Mol. Wt. 157.6, Sigma Cat No. T-3253
  23.38 g NaCl
  15.21 g EDTA, Anhydrous Mol. Wt. 380.2, Sigma Cat No. ED4SS
- Place flask on a magnetic stirrer until dissolved.
- Add additional water to make 1.0 liter and mix well.
- Store in the refrigerator.

Spermine

- Spermine tetrahydrochloride, Cat No. 5677.
- 13.93 g Spermine in 100 ml water.

Spermidine

- spermidine trihydrochloride, Cat. No. 56766.
- 25.46 g spermidine in 100 ml water.
- Spermine and spermadine are available from;
  Calbiochem, Behring Diagnostics,
  La Jolla, CA 92037.
- Spermine and spermidine are aliquoted into 2.0 ml portions and stored in the freezer. The required 2.0 ml aliquots of each are thawed just before use.

References

van den Engh G, Trask BJ, Cram LS, Bartholdi MF. Preparation of chromosome suspensions for flow Cytometry. Cytometry 1984; 5: 108.

van den Engh G, Trask BJ, Gray JW, Langlois RG, Yu L-C. Preparation and bivariate analysis of suspensions of human chromosomes. Cytometry 1985; 6: 92.

Silar R, Young BD. J. Histochem. Cytochem 1981; 29: 74–48.

Author for correspondence: L. Scott Cram, National Flow Cytometry Resource, Los Alamos National Laboratory, Los Alamos, New Mexico, 87545, USA

Methods in Cell Science **24**: 37–40 (2002).

Analysis of intracellular cytokines using flowcytometry

Sunil K. Arora
Department of Immunopathology, Postgraduate Institute of Medical Education & Research, Chandigarh-160 012, India

Abstract. Characterization of T-cell clones and identification of functional subsets of the helper T-cells with polarized cytokine production is based on testing of cytokine expression. Several methods have been developed that allow cytokine expression to be measured like ELISA, RT-PCR, ELISPOT, ISH and flowcytometry. Among all these methods, monitoring of cytokine production using flowcytometric analysis has its own advantages and disadvantages. Multi-parametric characterization of cytokine production on single cell basis, without long-term culture and cloning along with high throughput of samples is main feature attached to flowcytometric analysis. The interpretation may be difficult at times due to change in the phenotype of the cells. Cells with similar surface phenotype but synthesizing different cytokines and having different functional characteristics can be analyzed with this technique.

Key words: Cytokines, Flowcytometry, Intra-cellular

1. Introduction

Flowcytometry is a very refined and advanced technique in the field of immunology. Although, surface markers on some of the cells correlate with their function, it may not always be possible to infer the function of a cell by surface staining as phenotypically similar cells may perform different functions *in vivo* due to the expression of a different set of proteins intracellularly. This fact has been proven beyond doubt with the characterization of Th1 and Th2 subpopulations of CD4+ T-cells depending on the expression of different profile of cytokines by the two subsets. The identification of functional subsets of CD4+ T-lymphocytes with polarized cytokine production was based originally on the characterization of T-cell clones using cytokine ELISA of culture supernatants [7]. However, ELISA becomes impractical when large numbers of heterogeneous cell populations are to be analyzed *ex vivo*. It may be very difficult to develop enough number of clones because it is laborious as well as not possible in some of the cell populations [6].

There are various methods, which have been developed for measuring the expression of cytokines by the cells, these being-enzyme linked immunosorbant assay (ELISA), Enzyme-linked immuno-spot assay (Elispot), Reverse-transcription-polymerase chain reaction (RT-PCR), *in-situ* hybridization (ISH), immuno-histochemistry, LDA and intracellular cytokine staining (ICCS). There are advantages and drawbacks attached to each assay system. While ELISA measures total amount of secreted protein, the RT-PCR measures the levels of mRNA of a specific gene semi-quantitatively. The localization of cytokine producing cells in tissues can be achieved by immuno-histochemistry and *in-situ* hybridization. Flow cytometric analysis for intracellular cytoplasmic staining of cells allows individual characterization of large number of cells displaying the heterogeneous nature of cells completely. It is possible to characterize the cells on the basis of cytokine they express, rather then only surface markers, due to the possibility of multicolor staining that can demonstrate exclusive or mutual co-expression of different cytokines in individual cells.

Intracellular cytokine staining is relatively a new technique introduced initially for immuno-staining of tissue sections [5], but was quickly adopted for staining the intracellular cytokines in permeabilised and fixed cells in suspension with the availability of monoclonal antibodies and reagents for fixing and making the cell membranes permeable to the reagents [4]. Flowcytometric analysis of the intracellular cytokine stained cells allows characterization of large number of cells displaying the heterogeneity of various cell populations. The advantage over existing techniques involving enzyme immuno assays is that multicolor staining availability in intracellular cytokine staining demonstrates the co-expression of different cytokines in individual cells which allows the characterization of various sub-populations of cells on the basis of expression of the profile of cytokines along with the markers on their surface.

2. Cell source

Cytokines are the soluble factors released by many different types of cells as a sequel of activation in response to some stimulus. The release of cytokines is the measure of their function. Though various different types of cells release different set of cytokines, but it is more conventional to study the intracellular cytokine expression in lymphocyte suspensions and so more and more studies are available in this field only. Peripheral blood mononuclear cells (PBMC) separated by density gradient techniques are most commonly used for flowcytometric analysis, but methods using whole blood are also available. Though flowcytometry allows analysis of lymphocytes in more complex mixtures such as broncho alveolar lavage (BAL) without separation or culture, yet important histological information may be lost in the exfoliated cells.

3. Cell stimulation

Cells from the patients chronically infected with various infectious agents may be already activated and producing some or the other cytokine and so may be used directly for the analysis. But under the circumstances when the number of cells producing a particular cytokine is expected to be very small, some form of external activation may be required before the analysis could be done. Such activation is possible by using a specific antigen if that is known or by using non-specific stimulators like phorbol myristate acetate (PMA) and ionophore, phytohemagglutinin (PHA) or monoclonal antibodies such as anti-CD3 or anti-CD28 antibodies. The use of polyclonal activators is usually the only option with polyclonal responses. However, it should be ascertained that the cytokine profile is the same as being induced with the physiological stimulus because the polyclonal activators will induce a broader spectrum of cytokines production from the T-cell clones as compared to the specific antigen. The timing of the stimulation is also important as the T-cells enter a refractory phase after stimulation. Thus T-cell lines and clones maintained in culture are most responsive when their periodic stimulation is due.

Cell suspensions from various sources like the peripheral blood, BAL, lymphnode cells, thymocytes, splenocytes etc. are prepared in the medium are activated by specific antigens with antigen presenting cells or polyclonal stimulators, depending on the specificity of the cell source in short term cultures (4–6 hours). Block of cytokine secretion and intracellular accumulation is achieved by treatment with Brefeldin A (Golgi-plug) or monensin (Golgi-stop).

4. Cell fixation and permeabilization

The cells after activation will need to be fixed, which is best achieved by treatment with formaldehyde. Fixation with formaldehyde preserves the cytokine antigenecity and other characteristics of the cells without causing any artifacts. Clumping of cells due to fixative may be avoided by vortexing. The fixed cells are permeabilized using reagents like saponin, which allows the antibodies to enter the cytoplasm and react with the intracellularly accumulated cytokines.

5. Staining

The fixed and permeabilized cell suspensions are compatible with both direct and indirect staining. The direct staining with monoclonal antibody conjugates (from Pharmingen, Sigma, DNAX, Medprobe Biosource etc.) usually produce lower background and is simple, faster and easy for demonstration of co-expressed cytokines. Cell surface markers like CD3, CD4, CD16, CD56, CD19, CD8 or T1/ST2 (Th2 marker) very well correlate with various degrees of cytokine production. Surface staining is best performed before fixation because epitopes in surface markers may be destroyed by fixation and permeabilisation. However, it is not always possible to characterize the cells on the basis of their cell surface markers, as some of the surface markers are down-regulated due to stimulus used for cytokine secretion. For example the CD4 surface markers are affected severely by treatment with PMA and ionomycine, while such effect is of lower intensity on surface markers such as CD8, TCR and CD3. Thus, either it is possible to determine accurately the surface phenotype of a cell or the intracellular cytokine production, but not both at the same time leading to a sort of uncertainty. There are some modalities possible like use of high concentration of reagent that could inhibit the endosomal degradation e.g. monensin, but that may effect the expression of cytokine.

6. Controls

Appropriate positive and negative controls are very important for any technique in order to get a useful and accurate interpretation of results. The positive controls in intra-cellular cytokine staining can be activated and fixed cells that are commercially available or T-cell lines or clones with known cytokine profile may be used after stimulation with polyclonal activators. Since the procedure demands fixation and permeabilization of cells, these treatments lead to some artifacts. Fixation of cells causes non-specific trapping of antibodies leading to increased autofluorescence and the permeabilization increases the amount of protein available for non-specific interaction that can be avoided by careful titration of staining antibody. Blocking step may reduce the non-specific staining. Use of permeabilization controls like labeled anti-vimentin or anti-β-actin antibodies may be useful.

7. Flow cytometry

The flowcytometric analysis of the intracellular cytokine stained cells is not very much different in principle from the surface marker staining. Multicolor cytofluorimetry needs appropriate electronic compensation, which may be a little problematic as the number of colors increase. The treatment of cells with various reagents to fix or permeabilize the cells affects the scatter characteristics of the cells and so gating helps to confirm the identity of cell populations. It is always helpful and more reliable for analysis of expression of a cytokine to acquire large number of events in case dealing with a rare cell phenotype.

8. Pros and cons

The use of cytofluorometry has revolutionized the analysis of cell populations based on the surface markers and also the functional aspects of that cell depending on the expression of cytokine profile [3]. This has helped in defining the functional subsets in mixed T-cell populations, besides it also allows the frequency of the cells producing that cytokine and the levels of expression. The multicolor staining also helps to investigate the co-expression of different cytokines. However, there are various limitations and pitfalls attached to this technique. The permeabilisation causes high level of autofluorescence. There are many antibodies, which do not bind very well to the permeabilised and fixed cells. Moreover, it is very difficult to appreciate the expression of those cytokines, which are expressed at a very low level and do not appear as a peak in the histograms.

Then there are problems attached to sampling error and the inherent characteristics of some cell populations, as the cytokine synthesis by such cells is not constitutive and only a small population of cells obtained *ex vivo* show the presence of intracellular cytokines. This is mainly because of the fact that all the cells obtained may not express that cytokine synchronously and kinetics of expression for different cytokines may be different and so the optimal time of detection vary [4]. That limits the investigation of co-expression of more then one cytokine by the same cells. Though it is possible to determine the number of cells expressing more then one cytokine at a single point of time, yet the sequential expression of cytokines by the same cells may not be possible by this technique.

Some subpopulations of cells are very transiently available and the cytokine they produce are very faintly stained, e.g. Th2 cells producing IL4, whereas others like Th1 cells producing IFN-γ or cells producing TNF-α are present for a long period of time and are stained very intensely [3]. It may be misleading at times as the abundant cell type may not be always central to the pathogenesis of a response. Moreover, there are some surface markers which are down-regulated after stimulation, for example CD4 receptor on the surface of cells decrease in number as indicated by loss of staining after stimulation for the expression of IFN-γ, this may lead to a wrong conclusion that the IFN-γ is being produced by cells other then CD4+ cells like NK cells or CD8 cells, while the fact may be different. Further, the cells have to be killed for the analysis of intracellular staining of cytokines and so no more functional studies are possible on these cells, allowing only a static view of cytokine production. There are some recent advancements in this field where very sensitive surface staining techniques that do not require fixation like development of megnato-fluorscent liposomes to detect surface IFN-γ [2] and isolation of single living cells inside microdroplets of the porous biomatrix.

9. Applications

The intracellular cytokine technique is useful for the characterization of immune responses in normal and diseased states. In humans most people have used peripheral blood mononuclear cells, but other body fluids like synovial fluid, tissue infiltrating lymphocytes, eosinophils, basophils and dendritic cells have also been examined. Although the PBMC are easier to obtain then locally infiltrating cells, it is important to recognize that relevant cells may accumulate in inflamed tissues and be depleted from the circulation during acute inflammatory conditions. The PBMC may not contain cells that are relevant to the response. The frequency of specific cells may increase during convalescence [1]. New technical advancements in this area have made powerful additions to the techniques that allow us to follow the fate of differentially activated lymphocytes as they develop. Such and newer techniques may extend the applications of flowcytometery to measurement of cytokine production in near future.

References

1. Isaacs D, MacDonald NE, Bangham CRM, McMichael AJ, Higgins PG, Tyrrell DAJ (1991). The specific cytotoxic T-cell response of adult volunteers to infec-

tion with respiratory syncitial virus. Inmun Infect Dis 1: 5.

2. Assenmacher M, Scheffold A, Schnitz J, Checa JAS, Miltenyi S, Radbruch A. (1996). Specific expression of surface interferon gamma on T-cells from mouse and man. Eur J Immunol 26: 263.
3. Pala P, Hussell T, Openshaw PJM (2000) Flow cytometric measurement of intracellular cytokines. J Immunol Meth 243: 107–124.
4. Openshaw P, Murphy EE, Hosken NA, Maino V, Davis K, Murphy K, O'Garra A (1995). Heterogeniety of intracellular cytokine synthesis at the single cell level in polarized Th1 and Th2 populations. J Exp med 182: 1357.
5. Sander B, Andersson J, Andersson U (1991). Assessment of cytokines by immunofluorscence and the paraformaldehyde-saponin procedure. Immunol Rev 119: 65.
6. Lalvani A, Brookes R, Hambleton S, Britton WJ, Hill AV, McMichael AJ (1997). Rapid effector function in CD8+ memory T-cells. J Exp Med 186: 859.
7. Mosmann TR, Cherwinski HM, Bond MW, Geidlin MA, Coffman RL (1986) Two types of murine helper T-cell clone. I. Definition according to profiles of lymphokine activities and secreted proteins. J Immunol 136: 2348.

Address for correspondence: Dr. Sunil K. Arora, Department of Immunopathology, Postgraduate Institute of Medical Education & Research, Chandigarh-160 012, India
Fax: +91-172-744401 & 745078;
E-mail: skarora_in@yahoo.com

Methods in Cell Science 24: 41–47 (2002).

Cell array coupled with laser scanning cytometry allows easy analysis of changes in cyclin expression during the cell cycle

An application of cell array system

Tomoko Furuya[1], Morihito Takita[1], Shin-ichi Tsunoda[2], Shigeto Kawauchi[1], Takashi Hirano[2], Atsunori Oga[1] & Kohsuke Sasaki[1]

[1] *Department of Pathology, Yamaguchi University School of Medicine, Ube 755-8505, Japan;* [2] *Institute of Molecular and Cell Biology, National Institute of Advanced Industrial Science and Technology, Tsukuba 305-8566, Japan*

Abstract. To assess the potential of cell array technology, a cell array slide with 50 spots was designed specifically for automated analysis of changes in expression of cyclins A and B1 during the cell cycle at the cellular level. Cells harvested every 1 hour from 0 through 23 hours after synchronization by mitotic selection were spotted in duplicate on a glass slide. Each slide contained 48 spots representing 24 different cell cycle phases, and the remaining two spots were peripheral blood lymphocytes, which were included as negative control. This cell array provided temporal and spatial information related to changes in cyclin expression during the cell cycle in a single experiment. The present study indicates that expression analysis by cell array is novel approach for cell cycle studies. Furthermore, sophisticated multiplexed cell array technology has great potential for analyses of expression of specific genes during diverse cellular events at the cellular level.

Key words: Cell array, Cell cycle, Gene expression, Laser scanning cytometry (LSC), Synchronization

1. Introduction

In the post genome-sequencing era, one important subject of research is the function of each gene within a cell. It is important to understand when and where genes are turned on and off in a cell, because product of each gene influences cellular characteristics alone or in combination with other proteins. Qualitative and/or quantitative alterations in proteins may have pathological consequences. Functional analyses of individual genes and relational analyses of many genes are necessary and may provide insights into the normal cellular phenotype and into the developmental mechanisms of disease. Protein levels within cells are responsible for the biological state of the cell, and intracellular localization and kinetics of proteins also influence cell function [1, 2, 13]. Therefore, knowledge of both the amounts of proteins and their intracellular distributions in individual cells in multiple samples is essential. To meet these demands, development of an efficient method to analyze gene function at the cellular level is necessary.

We developed a novel high-throughput method called a cell array system to analyze characteristics at the cellular level and previously applied this technique to DNA ploidy and FISH analysis of tumor cell lines [10]. This technology has great potential for both basic researches, such as gene expression analysis, and clinical applications, such as tumor marker analysis. The cell array system including laser scanning cytometry (LSC) allows quantitative high-throughput analysis of gene expression at the cellular level [10].

Genes expressed in a cell cycle-dependent manner, such as cyclins, are thought to play important roles in cell proliferation. It is important to determine the timing of both expression of genes and degradation of the resulting proteins. Changes in the intracellular distributions of proteins during the cell cycle may also be informative as to the functions of such proteins. A specially designed cell array may provide detailed information regarding gene expression linked to particular cellular events such as the cell cycle.

A cell array slide spotted with multiple samples of cells harvested at constant intervals after synchronization permits a detailed analysis of changes in cyclin expression during the cell cycle in a single experiment. Furthermore, this system also provides information about intracellular distribution and kinetic patterns.

In this brief communication, we describe the use of the cell array system in an analysis of changes in cyclin expression associated with the cell cycle and discuss the potential impact of the cell array system in functional analyses of genes at the cellular level.

2. Materials

2.1. *Cell array slide glass*

As previously reported [10, 14], a glass slide was designed specifically for analysis of 50 samples in a single experiment (Figure 1). There are 50 (5 × 10) 2-mm diameter circles in a 16 × 31-mm square area which are covered with black printing ink without significant autofluorescence. The number (1 to 10) and letters (A to E) are printed along the margins of the printed area to easily identify the position of individual circle. The surface of circle is coated with triethoxyamino-propylsilane to prevent desquamation of cells from the slide glass.

2.2. *Cells*

We used HeLa S3 cells and human peripheral blood lymphocytes. HeLa cells were grown in Dulbecco's modified Eagle medium (Sigma Chemical Co., St. Louis, MO) supplemented with 10% calf serum at 37 °C in a humidified atmosphere of 5% CO_2 in air.

3. Procedures

3.1. *Cell synchronization*

The mitotic cell selection method was used for cell synchronization because there is a possibility that drugs perturbing the cell cycle may affect the expres-

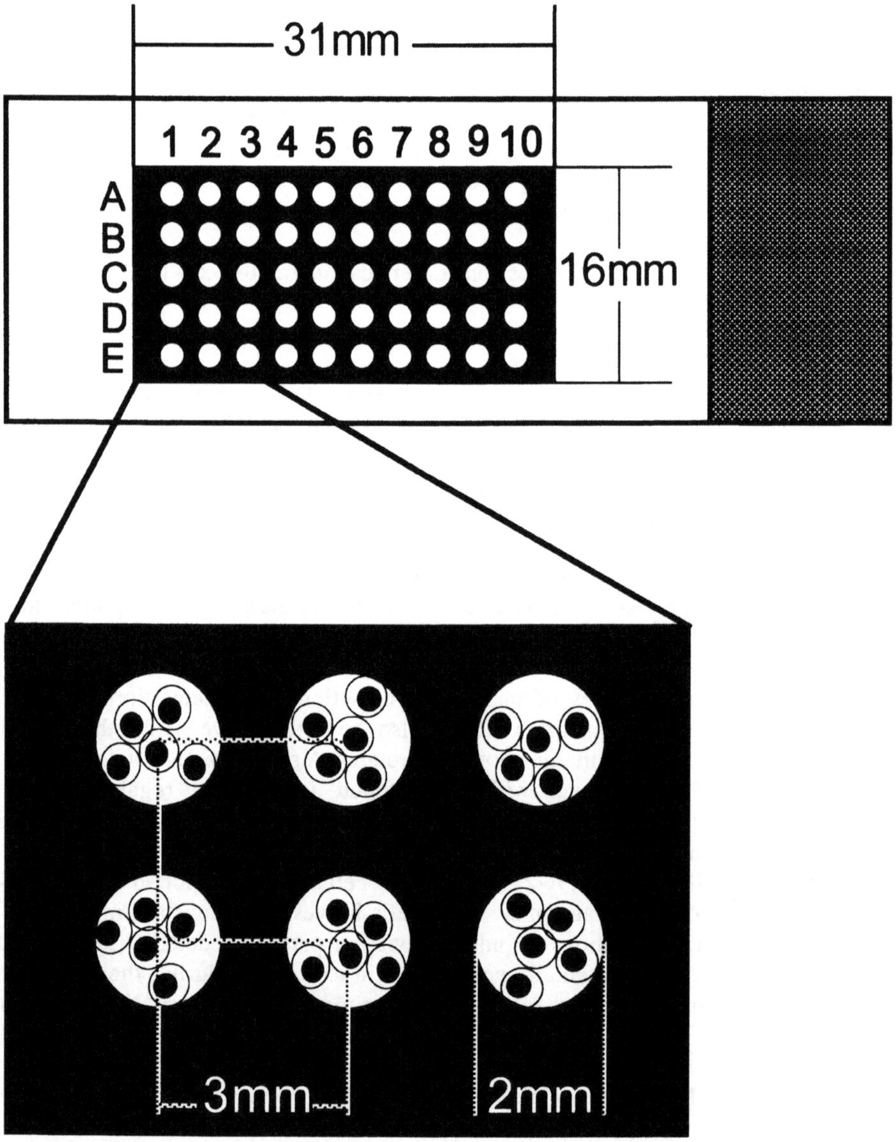

Figure 1. A specially designed device designated as a cell array slide glass. There are 50 (10 × 5) 2-mm diameter circles in a 30 × 16 mm area which is covered with black printing ink without significant autofluorescence. The circles are coated with triethoxyaminopropylsilane. The number (1 to 10) and letters (A to E) are printed along the margins of the printed area, allowing easy identification of the position of individual circle.

sion level of cellular components [15]. HeLa cells were synchronized by the mitotic cell selection technique and then cultured, as previously described [14].

1. Seed HeLa cells in 175-cm^2 flasks (Becton Dickinson Co., Franklin Lakes, NJ).
2. Change the culture medium to exclude cell debris and dead cells.
3. Tap the flasks vigorously several times.
4. Transfer the medium from the flasks to centrifuge tubes.
5. Centrifuge cells at 1,000 rpm for 5 min to collect cells.
6. Resuspend cells gently in the prewarmed culture medium.
7. Divide the medium containing cells into 24 dishes (19.6-cm^2, Becton Dickinson Co.) to culture the cells.

To measure the percentage of mitotic cells, a smear slide was prepared using a small amount of cell suspension.

8. Harvest cells with 0.25% trypsin in phosphate-buffered saline (PBS) at 1-hour intervals from 0 through 23 hours after mitotic selection.
9. Fix the cells in 70% ethanol immediately after harvest, and store them at –20 °C.
10. Isolate lymphocytes from peripheral blood by Ficoll-Hypaque (Amersham Pharmacia Biotech, Uppsala, Sweden), washed in PBS, fixed in 70% ethanol, and store them at –20 °C until use.
11. All cells were transferred into 100% ethanol immediately before cell array preparation.

The cell cycle progression was monitored by nuclear DNA content and cellular morphology. Twenty-four different cell cycle phases were examined for cyclins on the same slide. Because periods of the cell cycle (Tc), G1 (Tg1), DNA synthesis (Ts), and G2M (Tg2m) of HeLa cells were estimated to be 21, 9, 8, and 4 hours, respectively [11, 12], the cell array covered the entire length of the cell cycle.

3.2. *Cell array preparation for cell cycle analysis*

12. Put a cell array slide on a heating plate.
13. Put drops of fixed-cell suspensions in duplicate on the cell array glass slide with a specially designed chip to reduce variability among samples [10, 14].

Each cell array contains duplicate spots of 24 cell populations (2 × 24 spots). The number of cells in a spot is adjusted manually to be approximately 300 to 500 cells, because the high cell density in a spot results in overlapping cells and the aggregated cells are excluded from counting.

14. Drop peripheral blood lymphocytes in the remaining two circles as controls for the cultured cells.

3.3. *Immunocytochemical staining of cyclins*

An indirect immunofluorescence method is used to stain cyclins (in this study, use monoclonal antibodies against cyclins A and B1), as usual.

15. Incubate the cell array slides with normal goat serum diluted 1:10 in PBS for 30 minutes at room temperature to reduce undesirable nonspecific background staining.
16. Incubated for 2 hours at room temperature with monoclonal antibodies to cyclins A and B1 (clones 6E6 and 7A9, respectively, Novo Castra, Newcastle upon Tyne, UK), diluted 1:50 in PBS.
17. Wash the slides with PBS at least three times.
18. Incubate the array slides with Alexa Fluor 488-labeled anti-mouse IgG antibody (Molecular Probes, Inc., Eugene, OR) diluted 1:200 in PBS for 1 hour at room temperature.
19. Wash the slide in a PBS solution well.

3.4. *DNA staining*

20. Treat the slides with RNase (1 mg/ml, Sigma) for 30 minutes at room temperature to remove RNA from cells.
21. Transfer the slide to PBS solution containing propidium iodide (PI, 5 μg/ml, Calbiochem, La Jolla, CA).
22. Mount the slides in an anti-fade solution (Slow Fade, Molecular Probes) containing propidium iodide (PI, 5 μg/ml, Calbiochem).
23. Place coverslips on the slide.
24. Seal the slides with nail polish.

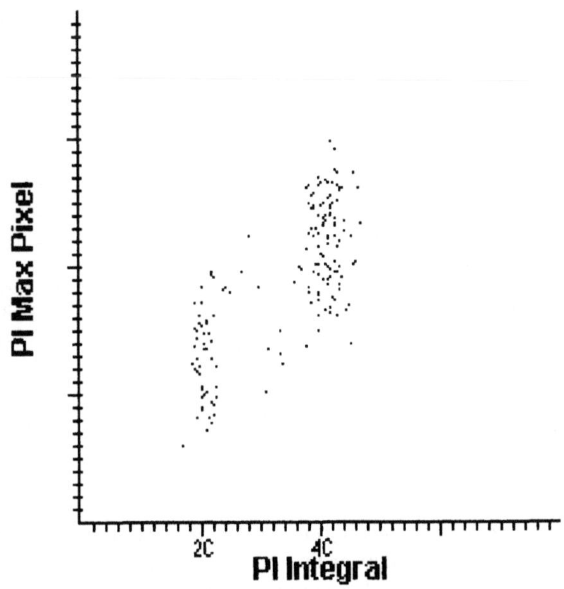

Figure 2. DNA cytograms indicate the cell cycle position of each cell. Most of cells are present in the mitotic phase or post-mitotic phase. Abscissa: PI fluorescence value representing DNA content, ordinate: peak fluorescence value representing the status of chromatin condensation.

(A)

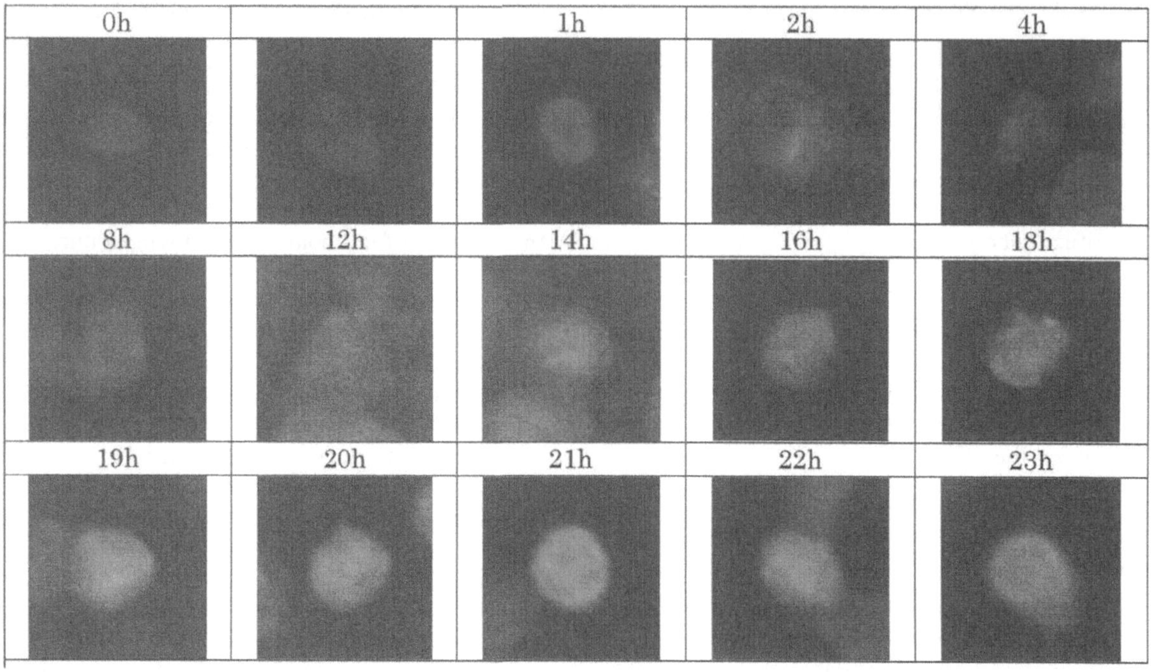

(B)

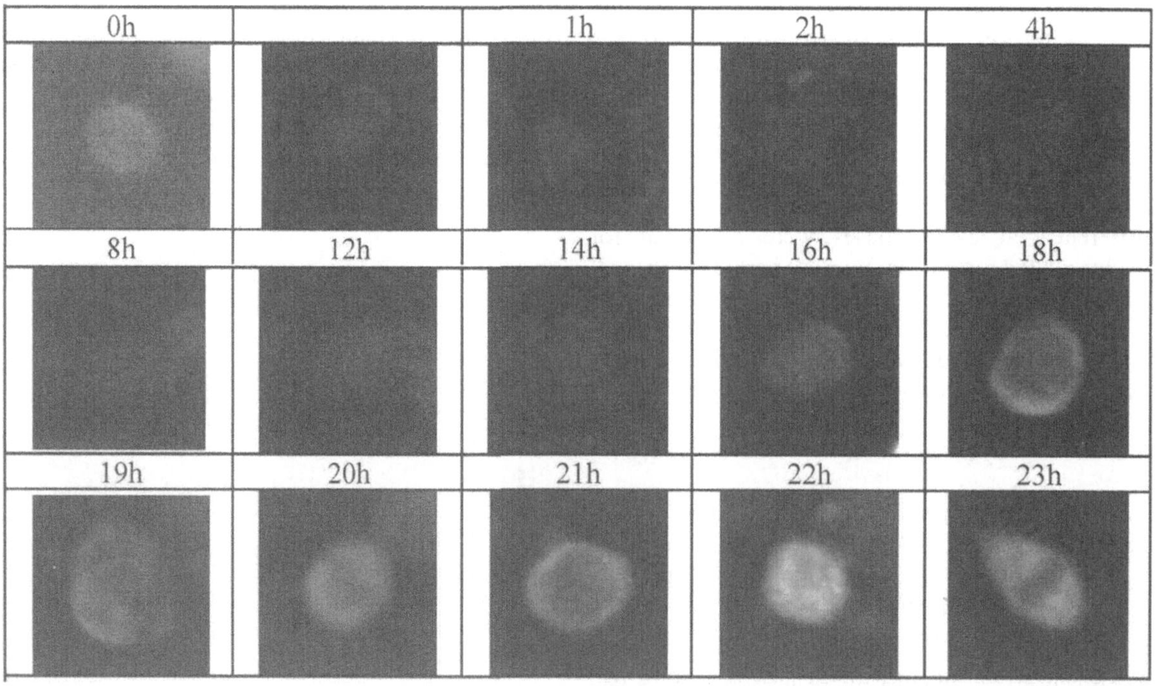

Figure 3. Changes in fluorescence images of cyclin expression during the cell cycle. A series of fluorescence images allows visual understanding of cell cycle phase dependent cyclin expression. A: cyclin A. B: Cyclin B1. Figures represent time (hours) after mitotic cell selection. Temporal course of cyclin expression during the cell cycle (14). C: cyclin A, D: cyclin B.

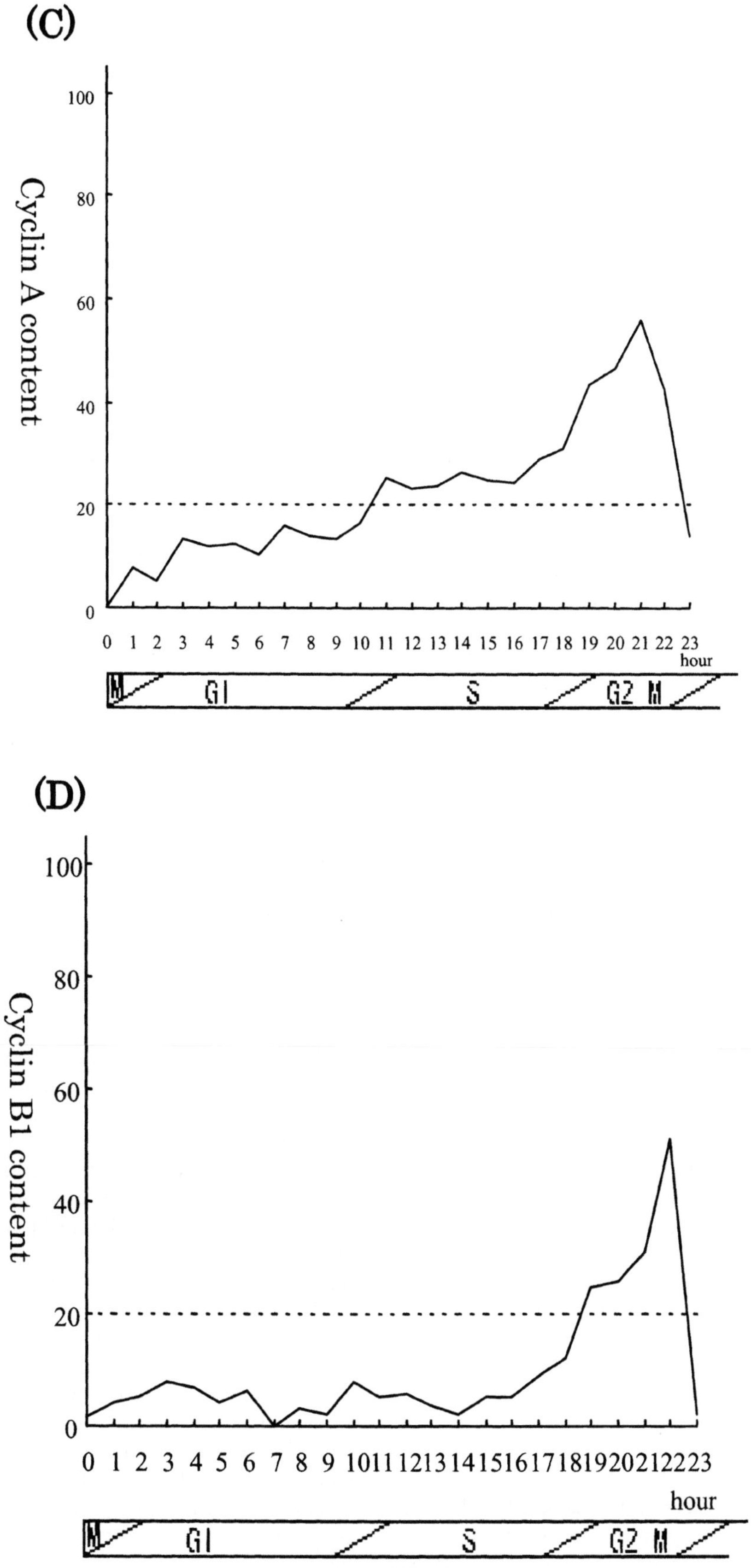

Figure 3. Continued.

25. Measure cyclin (cyclin A or B1) and DNA contents simultaneously with LSC as described below.

3.5. *LSC measurement of cyclins and nuclear DNA*

Amounts of cellular cyclins and nuclear DNA are measured concomitantly with a laser scanning cytometer (LSC2, CompuCyte, Cambridge, MA) for all spots at once. Scanning is limited to the area containing spots, and inter-spot areas are excluded from scanning by computer software. Fluorescence intensity of Alexa Fluor 488-labeled cyclins is measured in each cell. These parameters are stored together with cellular position data for all cells in list mode. Overlapping nuclei are excluded automatically from the analysis with a special statistical filter [4]. The cell array allows any combination of spots to be grouped and evaluated. A cyclin content histogram for each spot is constructed by setting a gate window on an X-Y scattergram that represents the distribution of spots. The intracellular cyclin content is represented by the mode values of fluorescence intensity for a cell population in each spot of the cell array. In addition, fluorescence images of cells are observed and photographed.

4. Results

In total, approximately 12,000 cells in 50 spots were examined for cyclin expression and nuclear DNA content by LSC for each experiment.

4.1. *Cell synchronization*

On the basis of both cellular morphology and LSC cytograms, approximately 80% of the cells obtained with mitotic selection were estimated to be in the mitotic and post-mitotic phases (Figure 2). Almost all cells obtained with the mitotic selection method were in G1 phase after 1 hour. The synchronization gradually distorted with time, however, this was not a serious hindrance to this study. LSC allows us to distinguish mitotic cells and post-mitotic cells from G2 cells and G1 cells, respectively, because the chromatin condensation is clearly different between them [5, 7]

4.2. *Temporal course of cyclin expression during the cell cycle*

Cyclin A
No significant fluorescence was detected in the cells 1 to 10 hours (corresponding to G1 phase) after mitotic selection (Figure 3-A). Cyclin A was detected in the nuclei of cells harvested 12 to 23 hours after mitotic selection, but significant amounts of cyclin A were not present in the cytoplasm. Cyclin A was expressed at high levels 19 to 22 hours after synchronization (corresponding to G2 phase). The highest fluorescence intensity was observed in a sample obtained 21 hours after mitotic selection (corresponding to G2 phase) (Figure 3-A).

Cyclin B1
Cyclin B1 was expressed scarcely in G1 and early S phase cells, but faint expression was observed16 hours after mitotic selection (corresponding to mid S phase). Cyclin B1 then accumulated rapidly in cytoplasm during G2 (16 to 20 hours after mitotic selection), reached a maximum 22 hours after mitotic selection (corresponding to early mitotic phase), and decreased to baseline levels by 23 hours after mitotic selection (corresponding to late mitotic phase) (Figure 3-B). Cyclin B1 was localized exclusively in the cytoplasm in G2 phase, but it was localized with the nucleus when the cell progressed through the mitotic phase (Figure 3-B).

5. Discussion

It is important to clarify the functions of genes to better understand cellular events. This requires high-throughput methods that allow analysis of cellular events coupled with expression of individual genes in a given cell. A method that provides information on both the spatial distribution and quantity of gene products in each cell would be very useful for functional analyses of genes. The cell array system including LSC is such a method [10, 14]. In the present study, the cell array system allowed us to examine temporal changes in cyclin expression during the cell cycle in a single experiment. The results with this cell array system were consistent with data obtained with other methods [1, 2, 8, 9, 13]. The cell array system also permitted observation of the spatial distributions of cyclins within a cell during the cell cycle. The use of cell in suspensions makes it easy to replicate cell array slides, and thus the cell array system can be used for expression analyses of multiple genes in a parallel fashion. To check up the specificity, sensitivity and reproducibility of each analysis, each cell array slide must contain the same control cells.

A tissue microarray enables analysis of protein expression in hundreds of tissue specimens in a single experiment [3, 6]. Although cell array and tissue microarray technologies are both based on detection of intracellular antigen, there are significant differences between the two technologies. For example, in cell array system, cell array slides can be flexibly arranged for the purpose of the study because cell lines can be used. In the present study cell array slides composed of synchronized cells harvested at constant intervals were used to study changes in cyclin expression during the cell cycle.

Such slides, which allow multiplex measurements of gene products, provide detailed and reliable data in a single experiment. Although the cell array is superior to tissue microarray in the quantitative measurement of cellular constituents, it provides no histologic information. Therefore cell arrays can compensate for weakness of tissue microarrays, and vice versa.

Speed of measurement is an important issue with respect to analysis with the cell array system, and it depends primarily on scanning areas with LSC. The analytical speed in this study was improved greatly by reducing the scanning area as compared to that in a previous report [10]. The majority of time and effort in a cell array experiment is spent in manual spotting of cells on the array, which is a major obstacle to the broad use of cell array technology.

Acknowledgments

This work was supported in part by the New Energy and Industrial Technology Development Organization (NEDO) of Japan and by Grants-in-Aid for Scientific Research (Nos.#12557019 and 13877027) from the Ministry of Education, Science, Sports and Culture of Japan.

Notes on suppliers

Cell array slide: Not commercially available. However, Matsunami Glass Ind. Ltd., Kishiwada, Osaka, Japan, may provide the specially designed cell array slide as described here.

Laser scanning cytometer™ (LSC2): CompuCyte Co., Cambridge, MA, USA

Culture flask (175-cm^2, #3028): Becton Dickinson Co., Franklin Lakes, NJ, USA.

Culture dish (19.6-cm^2, #3002J): Becton Dickinson Co., Franklin Lakes, NJ, USA.

Ficoll-Hypaque (17-1440-02): Amersham Pharmacia Biotech, Uppsala, Sweden

Dulbecco's modified Eagle medium: Sigma Chemical Co., St. Louis, MO, USA

Trypsin (T-4799): Sigma Chemical Co., St. Louis, MO, USA

Propidium iodide (537059): Calbiochem, La Jolla, CA, USA

Anticyclins A (clones 6E6): Novo Castra, Newcastle upon Tyne, UK

Anticyclin B1 (clone 7A9): Novo Castra, Newcastle upon Tyne, UK

Alexa Fluor 488-labeled anti-mouse IgG antibody (A-11017): Molecular Probes, Inc., Eugene, OR, USA

RNase (R4875): Sigma Chemical Co., St. Louis, MO, USA

Anti-fade solution: Slow Fade (S-2828): Molecular Probes, Eugene, OR

References

1. Darzynkiewicz Z, Gong J, Juan G, Ardelt B, Traganos F (1996). Cytometry of cyclin proteins. Cytometry 25: 1–13.
2. Kakino S, Sasaki K, Kurose A, Ito H (1996). Intracellular localization of cyclin B1 during the cell cycle in glioma cells. Cytometry 24: 49–54.
3. Kallioniemi O-P, Wagner U, Kononen J, Sauter G (2001). Tissue microarray technology for high-throughput molecular profiling of cancer. Human Mol Genet 10: 657–662.
4. Kamentsky LA, Burger DE, Gershman RJ, Kamentsky LD, Luther E (1997). Slide-based laser scanning cytometry. Acta Cytol 41: 123–143.
5. Kawasaki M, Sasaki K, Satoh T, Kurose A, Kamada T, Furuya T, Murakami T, Todoroki T (1997). Laser scanning cytometry (LSC) allows detailed analysis of the cell cycle in PI stained human fibroblasts (TIG-7). Cell Prolif 30 (3–4): 139–147.
6. Kononen J, Bubendorf L, Kallioniemi A, Barlund M, Schraml P, Leighton S, Torhorst J, Mihatsch MJ, Sauter G, Kallioniemi OP (1998). Tissue microarrays for high-throughput molecular profiling of tumor specimens. Nat Med 4: 844–847.
7. Luther E, Kamentsky LA (1996). Resolution of mitotic cells using laser scanning cytometry. Cytometry 23: 272–278
8. Mait A, McKenna WG, Muschel RJ (1997). Cyclin A message stability vaires with the cell cycle. Cell Growth Differ 8: 311–318.
9. Ohtsubo M, Theodoras AM, Schumacher J, Roberts JM, Pagano M (1995). Human cyclin E, a nuclear protein essential for the G1-to-S phase transition. Mol Bell Biol 15: 2612–2624.
10. Oode K, Furuya T, Harada K, Kawauchi S, Yamamoto K, Hirano T, Sasaki K (2000). The development of a cell array and its combination with laser scanning cytometry allows a high-throughput analysis of nuclear DNA content. Am J Pathol 157: 723–728.
11. Sasaki K, Murakami T, Ogino T, Takahashi M, Kawasaki S (1986). Flow cytometric estimation of cell cycle parameters using a monoclonal antiboy to bromodeoxyuridine. Cytometry 7: 391–395.
12. Sasaki K, Murakami T, Takahashi M (1987). A rapid and simple estimation of cell cycle parameters by continuous labeling with bromodeoxyuridine. Cytometry 8: 526–528.
13. Sherwood SW, Rush DF, Kung AL, Schimke RT (1994). Cyclin B1 expression in HeLa S3 cells studied by flow cytometry. Exp cell Res 211: 275–281.
14. Takita M, Furuya T, Sugita T, Kawauchi S, Oga A, Hirano T, Tsunoda S, Sasaki K. An analysis of changes in the expression of cyclins A and B1by the cell array system during the cell cycle: Comparison between cell synchronization methods. In submission
15. Terashima T, Tolmach LJ (1961). Changes in X-ray sensitivity of HeLa cells during the division cycle. Nature 190: 1210–1211.

Address for correspondence: Kohsuke Sasaki, M.D., Department of Pathology, Yamaguchi University School of Medicine, Ube 755-8505, Japan
Phone: +81-836-22-2222; Fax: +81-836-22-2223
E-mail: kohsuke@po.cc.yamaguchi-u.ac.jp)

Methods in Cell Science **24**: 49–54 (2002).

Flow cytometry and its application in small animal oncology

Katja Culmsee* & Ingo Nolte
Clinic of Small Animals, School of Veterinary Medicine Hannover, Bischofsholer Damm 15, 30173 Hannover, Germany

Abstract. Flow cytometry measures multiple characteristics of single cells. The use of flow cytometry in the veterinary clinical laboratory has increased considerably during the past decade. The most common applications of flow cytometry in small animal oncology are measurement of DNA content in tumours and immunophenotyping of haematopoietic malignancies. DNA ploidy and S-phase rate of various tumours in dogs have been found to be independent predictor of patient outcome. In dogs with lymphomas immunophenotyping is recommended as a part of the patient work-up. Flow cytometry has shown to be a suitable method for immunophenotyping of canine lymphomas. However, it has not become a routine technique in small animal oncology yet. This report reviews the applications of flow cytometry in small animal oncology. Besides basic principles and technical aspects, the clinical relevance of DNA-analysis and immunophenotyping are discussed.

Key words: DNA-analysis, Dog, Flow cytometry, Immunophenotyping, Oncology

Abbreviations: BrdU = bromodeoxyuridine; CLL = chronic lymphocytic leukemias; CRBC = chicken red blood cells; CV = coefficient of variation; DI = DNA-index; LI = labelling index; T_c = cell cycle time; T_{pot} = potential doubling time

1. Introduction

Flow cytometry enables rapid analysis of multiple characteristics of single cells. Cell properties that can be measured include cell size, cytoplasmic complexity, DNA or RNA content, and a wide range of membrane-bound and intracellular proteins.

An important feature of flow cytometric analysis are that measurements is made separately on each cell. The applications of flow cytometry and cell sorting are numerous and in the last decade flow cytometry has been evolved to an essential component of veterinary clinical laboratories [41]. In small animal oncology most common applications are measurement of DNA content in tumours and immunophenotyping of haematopoietic malignancies in dogs. Flow cytometric DNA analysis has been found of value in determining the biologic behaviour of various tumours and predicting clinical outcomes. Although currently the value of flow cytometric DNA analysis in the diagnosis of tumours in dogs is very limited.

Besides DNA analysis immunophenotyping of haematopoietic malignancies is another important application of flow cytometry. In man flow cytometry has become the preferred method for the lineage assignment and maturational analysis of malignant cells in acute leukaemias and lymphomas. Although in haematopoietic malignancies of companion animals, except canine lymphoma, assessment of immunophenotype is rarely performed the utility of flow cytometric immunophenotyping has been demonstrated.

This report reviews proven applications of flow cytometry in small animal oncology. Besides basic principles and technical aspects, the clinical relevance of DNA-analysis and immunophenotyping are discussed.

2. Analysis of DNA

Flow cytometry has been used to analyse the DNA content in canine tumours since the beginning of the eighties [21].

Informations obtained by flow cytometric DNA analysis include the presence or absence of a DNA aneuploid cell population, the degree of its aneuploidy and the proliferative capacity of the tumour cells.

For the measurement of DNA content by flow cytometry fluorescent dyes that intercalate into the DNA helical structure in a stochiometric fashion are used. The fluorescent signal is directly proportional to the amount of DNA in the cell. For each sample about 10000 to 20000 cells are measured. The data generated by the flow cytometer is complied into a histogram that represents the cell cycle.

3. DNA-ploidy

The DNA content of tumour samples can either be diploid, aneuploid or pseudodiploid, which will not be detected by flow cytometry. The degree of DNA aneuploidy of a cell population is referred to as DNA index (DI). The DI represents the ratio of fluorescence intensity of the G_0/G_1 test population to that of a normal diploid cell population in the same specimen [20]. When aneuploidy is measured as a change in DNA content, as opposed to a change in chromosome it should be called as DNA aneuploidy [10].

Several studies evaluating the DNA content in canine tumours including mast cell tumours [1], mammary tumours [17, 18, 26, 29], transitional cell carcinoma [5], plasma cell tumours [13], lymphoma [31], osteosarcomas [12], thyroid carcinomas [36] and melanomas [4] have been published. Although DNA aneuploidy was a relatively common finding in malignant tumours [17, 26, 29], it was also found at various percentages in benign tumours [13, 17, 26, 29], mammary dysplasia [17] and in hyperplasia/inflammation of the bladder [5]. According to some reports [17, 21, 26] aneuploidy indicates possible premalignancies.

4. S-phase fraction

The proportion of S-phase cells obtained from a DNA-distribution curve gives an approximate indication of the proliferative rate. In dogs assessment of S-phase fraction has been performed in lymphomas [31], osteosarcomas [12] and mammary tumours [17, 18]. In the study analysing canine lymphomas the S-phase fraction varied from 4 to 37% [31] whereas in primary mammary tumours S-phase fractions ranging from 2 to 19% have been reported [26].

5. Potential doubling time and cell cycle time

Cell proliferation *in vivo* has been studied extensively by using various methods. While static parameters such as S-phase fraction or Ki-67 index can be informative, they give no indication of the rates of cell-cycle progression.

Information about the overall rate of tumour cell production can be achieved by determining the potential doubling time (T_{pot}) and the cell cycle time (T_c). The T_{pot} is defined as the expected doubling time of the tumour in the absence of cell loss and the T_c as the time interval within which one cell completes a mitotic cycle. T_{pot} and T_c of a tumour can be measured by flow cytometry using bromodeoxyuridine (BrdU). BrdU is incorporated into newly synthesized DNA and can be recognized by flow cytometry using commercially available monoclonal antibodies [9, 14]. Parameters which were evaluated include the labelling index (LI), that means the proportion of cells which take up BrdU, the T_{pot}, and the T_c.

BrdU *in vivo* labelling, performed by the method described by Begg and co-workers [2] has been used for evaluation of cell kinetics in different canine [22, 30, 34] and feline tumours [30]. After labelling with BrdU by a single injection *in vivo*, samples were taken 4 to 8 h thereafter and stained after fixation in ethanol. In dogs and cats BrdU, dissolved in sterile saline was given at a dose of 500 mg [22, 30, 34] and 100 mg [30], respectively, as intravenous injection. In none of the studies using BrdU *in vivo* labelling in dogs and cats any side effect of BrdU was reported.

In a study evaluating the impact of heterogeneity in the predictive value of kinetic parameters in canine osteosarcomas the variation between tumours of different dogs were significantly higher than intratumorale variation of T_{pot} and T_c. Obtaining of at least two samples in large tumours known to be heterogeneous has been recommended [22].

6. Technical considerations

Biological and technical parameters can influence flow cytometric DNA analysis. An important factor in determining the precision of a flow cytometer is the coefficient of variation (CV) across the peak of DNA from cells in G_1-phase. The smaller the CV, the more accurate the estimation of ploidy and S-phase fraction.

Various preparative methods, described elsewhere [16, 37] have been used for the DNA analysis of canine and feline tumours. Measurement of DNA content in fresh frozen tissue [17, 18, 26, 29, 31] was done according to the method published by Vindelov et al. [37]. For the estimation of DNA content in formalin-fixed, paraffin embedded tissue the method described by Hedley et al. [16] has been widely used [1, 4, 5, 13].

In different studies performed on paraffin-embedded tissue, a minor part of the canine tumours processed for flow cytometry did not have interpretable DNA histograms [4, 5]. In the majority of these tumours, the CV of the G_0/G_1 peak in the DNA-histogram was too high to assure appropriate data analysis [4, 5]. In one study problems (CV > 7.5) to interpret the results were encountered particularly in analysing heavily pigmented melanomas [4]. In assumption of the authors the opacity of the cell suspension that was seen after sample preparation in heavily pigmented melanomas quenched the delicate red nuclear fluorescence, displacing the peaks [4].

For the correct estimation of DNA-ploidy a standard, which is subjected to the same process as the sample is needed. In studies performed on canine fresh frozen tissue chicken red blood cells (CRBC) [17, 18, 29, 31] canine kidney cells [26, 29], canine lymphocytes [12] or rainbow trout red blood cells [29] can be used as internal standard. Additional to the internal standard an external standard, as such as CRBC [26] or canine peripheral blood nucleated cells combined with CRBC [17, 18] can be utilized.

For paraffin-embedded tissues, the use of external diploid controls is not recommended [16]. For this reason in the majority of the studies performed on paraffin embedded canine tissue normal diploid cells of tumour samples serve as internal standard [5, 13, 29].

7. Clinical relevance

Several studies evaluating the value of cellular DNA content in predicting the behaviour of canine tumours have been published until now.

Relatively high percentages of DNA aneuploid tumours have been reported in canine transitional cell carcinoma (79%) [5], thyroid carcinomas (58%) [36], osteosarcomas [12] and malignant mammary tumours (48 to 62%) [17, 26, 29]. Contrasted with this, the percentages of aneuploid tumours that have been found among canine mast cell tumours (28%) [1] and lymphomas (21%) [31] were lower.

In the majority of the tumours examined including canine transitional cell carcinomas [5] and canine lymphomas [31], no correlation between DNA ploidy and clinicopathological features was found.

In canine mammary tumours, DNA-aneuploid lesions tended to behave more aggressively than histologically and clinically similar lesions which were DNA-diploid [18]. Also a significant association between DNA-ploidy of canine primary mammary tumour and of their metastases has been reported [19].

In canine mast cell tumours a tendency towards an increased survival within the first years in dogs with DNA diploid versus DNA aneuploid tumours was determined [1].

However, no correlation was found between S-phase fraction and survival in a series of 50 dogs with malignant lymphoma treated with a combination chemotherapy protocol [31], T_{pot} was predictive of first remission duration in canine lymphoma in another study [34].

8. Immunophenotyping of leukaemias and lymphomas

In the last decade flow cytometry has evolved to an indispensable tool in the diagnosis of haematopoietic malignancies in man. It is commonly used for lineage assignment, analysis of cellular maturation and heterogeneity within the malignant cell population as well as detection of minimal residual disease and monitoring of disease [27]. In comparison to humane medicine the level of complexity, sophistication and importance which immunophenotyping has achieved in veterinary medicine is far lower.

Flow cytometric immunophenotyping is based on the identification of single or multiple surface membrane antigens or intracellular antigens of cells in suspension simultaneously with the light scatter characteristics of these cells. The antigens are detected with fluoresceinated monoclonal antibodies. Non-nucleated cells are depleted before analysis [27].

Although in 1994 the First International Canine Leucocyte Antigen Workshop identified a variety of canine specific antibodies that reacted with leucocytes, particularly lymphocytes [6], the immunophenotyping of haematopoietic malignancies in both the dog and cat is currently hampered by the paucity of specific antibodies. Nevertheless flow cytometry can be regarded as a promising tool in the diagnosis of haematopoietic malignancies in dogs and cats.

9. Leukaemias

Leukaemia is a relatively rare disorder in dogs and cats. In both species lymphoid leukaemia is assumed to be more common than myeloid leukaemia.

A panel of antibodies including anti-CD3, anti-CD4, anti-CD8, anti-CD5, anti-CD21, and anti-CD79a have been used to immunophenotype lymphoid leukaemias in dogs. According to their immunophenotype, the categorization of lymphoid leukaemias in lymphoproliferative malignancies of T-cell- and of B-cell-origin is attempted.

In a study on a large series of leukaemias in dogs, 73% of chronic lymphocytic leukaemias (CLL) were found to be of T-cell origin [35]. Similar results have been reported in other studies analysing fewer cases of CLL [7, 28]. It is assumed that in contrast to man [3], canine CLL is predominantly a neoplasia of T-cells [35].

Also in a recent study marked variations in the morphology and immunophenotype of canine acute lymphocytic leukaemia have been reported [39]. Of five dogs evaluated, two were categorized as T-cell prolymphocytic leukaemia, one as acute B-cell lymphoblastic leukaemia and two as acute non B lymphoblastic leukaemia [1]. CD34 expression, a common finding in acute, non large granular lymphocyte leukaemias of dogs was not detectable in CLL [35].

Flow cytometric scatter plot analysis is assumed to be a useful technique for evaluation of the lineage of leukaemic cells in blood and bone marrow [25, 39]. Scatter plot analysis correctly predicted the lineage of leukaemic cells in 7 of 8 dogs [11].

In another study, the combination of flow cytometric scatter plot analysis and four monoclonal antibodies (anti-CD18, anti-Thy-1, Anti-MHC class II, and anti-CD14) was used to identify the lineage of immature myeloid cells in six dogs with bone marrow proliferative disorders. Although the phenotype from myelogenous and myelomonocytic leukaemia clearly differed, myelomonocytic leukaemia could not be differentiated from malignant histiocytosis and haemophagocytic syndrome [40].

10. Lymphomas

In comparison to leukaemia assessment of immunophenotype is relatively widespread in canine lymphomas, because of its prognostic significance [15, 32]. Although the utility of flow cytometric immunophenotyping in canine lymphomas has been shown [8, 15, 28], immunohistochemistry is currently more popular. In one of the first studies using flow cytometry for the immunophenotyping of canine lymphomas, single-colour immunofluorescence analysis was performed on cryopreserved cell suspensions obtained from excisional lymph node biopsy specimens [15]. In a subsequent study two- and three colour immunofluorescence analysis was used on native cell suspensions taken by fine needle aspiration [8].

11. Technical considerations

Clinical application of immunophenotyping in canine and feline haematologic malignancies is currently hampered by the lack of systematic quality control procedures. In contrast to humane medicine [27] a consensus protocol for the flow cytometric immunophenotyping in haematopoietic malignancies of dogs and cats has not been established yet. For depletion of non-nucleated cells before analysis erythrocyte lyse procedures have been used [35, 39]. Although rubriblasts and rubricytes are not lysed by this procedure [38].

Flow cytometric analysis in dogs with leukaemia has been performed on bone marrow [39] as well as peripheral blood [28, 35]. Because aleukaemic forms of acute leukaemias are seen in dogs, bone marrow evaluation appears to be more sensitive than blood evaluation [39].

12. Clinical relevance

Malignant lymphomas are the most common haematopoietic malignancies in dogs [23]. B-cell lymphomas have been found to be more common than T-cell-lymphomas. Overall, the percentage of T-cell lymphomas reported in dogs ranged from 19% to 38% [8, 15, 33, 34].

Chemotherapy is the most effective treatment for lymphoma in the dog [23]. Although induction of remission is achieved in the majority (75–95%) of dogs receiving chemotherapy, and survival times often exceed 1 year, cure remains elusive [24]. In several studies the value of immunophenotype in predicting the success of chemotherapy in dogs with lymphoma has been shown. Median survival times and duration of first remission in dogs with B-cell-lymphomas were significantly longer than in dogs with T-cell-lymphomas [15, 32].

13. Summary

Flow cytometry is a powerful technique for analysing multiple characteristics on single cells. Applications of flow cytometry in small animal oncology include DNA analysis and immunophenotyping of leukaemias and lymphomas. Although in various canine tumours DNA-ploidy, S-phase fraction ant T_{pot} measured by flow cytometry have emerged as independent predictors for patient outcome, flow cytometric DNA-analysis has not become routine in veterinary clinical laboratories yet. Unfortunately, flow cytometers are not simple instruments. The flow cytometric equipment costs are comparably high and trained personnel is required. Nevertheless the clinical application of flow cytometry is advancing our knowledge of tumours in dogs.

References

1. Ayl RD, Couto CG, Hammer AS, Weisbrode S, Ericson JG, Mathes L (1992). Correlation of DNA ploidy to tumor histologic grade, clinical variables, and survival in dogs with mast cell tumors. Vet Pathol 29: 386–390.
2. Begg AC, McNally NJ, Shrieve DC, Karcher H (1985). A method to measure the duration of the DNA synthesis and the potential doubling time from a single sample. Cytometry 6: 620–626.
3. Bennett JM, Catovsky D, Daniel MT, Flandrin G, Galton DA, Gralnick HR, Sultan C (1989). Proposals for the classification of chronic (mature) B and T lymphoid leukaemias. French-American-British (FAB) Cooperation Group. J Clin Pathol 42: 567–584.
4. Bolon B, Calderwood Mays MB, Hall BJ (1990). Characteristics of canine melanomas and comparison of histology and DNA ploidy to their biologic behavior. Vet Pathol 27; 96–102.
5. Clemo FAS, DeNicola DB, Carlton WW, Morrison WB, Walker E (1994). Flow cytometric DNA ploidy analysis in canine transitional cell carcinoma of urinary bladders. Vet Pathol 31: 207–215.
6. Cobbold S, Metcalfe S (1994). Monoclonal antibodies that define canine homologues of human CD antigens: Summary of the First International Canine Leukocyte Antigen Workshop (CLAW). Tissue Antigens 43: 137–154.
7. Culmsee K, Mischke R, Nolte I (2001). Durch-

flusszytometrische Untersuchung von Knochenmark leukämiekranker Hunde. Tierärztl Prax 29: 176–181.
8. Culmsee K, Simon D, Mischke R, Nolte I (2001). Possibilities of flow cytometry for immunophenotypic characterization of canine lymphoma. J Vet Med 47: 199–206.
9. Dolbeare FA, Gratzner HG, Pallavicini MG, Gray JW (1983). Flow cytometric measurement of total DNA content and incorporated bromodeoxyuridine. Proc Natl Acad Sci USA 80: 5573–5577.
10. Dressler LG (1990). Controls, standards and histogram interpretation in DNA flow cytometry. Methods Cell Biol 33: 157–171.
11. Fernandes P, Modino J, Johnson M (1998). Determination of leukemic blast cell lineage of dogs and cats by flow cytometric light scatter. Vet Clin Pathol 27: 130.
12. Fox MH, Armstrong LW, Withrow SJ, Powers BE, LaRue SM, Straw RC, Gillette EL (1990). Comparison of DNA aneuploidy of primary and metastatic spontaneous canine osteosarcomas. Cancer Res 50: 6176–6178.
13. Frazier KS, Hines II ME, Hurvitz AI, Robinson PG, Herron AJ (1993). Analysis of DNA aneuploidy and c-myc oncoprotein content of canine plasma cell tumors using flow cytometry. Vet Pathol 30: 505–511.
14. Gratzner HG (1982). Monoclonal antibody to 5-bromo- and 5-iododeoxyuridine: a new reagent for detection of DNA replication. Science 218: 474–475.
15. Greenlee PG, Filippa DA, Quimby FW, Patnaik AK, Calvano SE, Matus RE, Kimmel M, Hurvitz AI, Lieberman PH (1990). Lymphomas in dogs. A morphologic, immunologic, and clinical study. Cancer 66: 480–490.
16. Hedley DW, Friedlander ML, Taylor IW, Rugg CA, Musgrove EA (1983). Method for analysis of cellular DNA content of paraffin-embedded pathological material using flow cytometry. J Histochem Cytochem 31: 1333–1335.
17. Hellmen E, Lindgren A, Linell F, Matsson P, Nilsson A (1988). Comparison of histology and clinical variables to DNA ploidy in canine mammary tumors. Vet Pathol 25: 219–226.
18. Hellmen E, Bergström R, Holmberg L, Spangberg IB, Hansson K, Lindgren A (1993). Prognostic factors in canine mammary tumors: A multivariate study of 202 consecutive cases. Vet Pathol 30: 20–27
19. Hellmen E, Svensson S (1995). Progression of canine mammary tumours as reflected by DNA ploidy in primary tumours and their metastases. J Comp Pathol 113: 327–342
20. Hiddemann W, Schumann J, Andreef M, Barlogie B, Herman CJ, Leif RC, Maball BH, Murphy RF, Sandberg AA (1984). Convention on nomenclature for DNA cytometry. Committee on Nomenclature, Society for Analytical Cytology. Cancer Genet Cytogenet 13: 181–183.
21. Johnson TS, Raju MR, Giltinan RK, Gillette RL (1981). Ploidy and DNA distribution analysis of spontaneous dog tumors by flow cytometry. Cancer Res 41: 3005–3009.
22. LaRue SM, Fox MH, Withrow SJ, Powers BE, Straw RC, Cote IM, Gillette EL (1994). Impact of heterogeneity in the predictive value of kinetic parameters in canine osteosarcomas. Cancer Res 54: 3916–3921.
23. Leifer CE, Matus RE (1986). Canine lymphoma: Clinical considerations. Sem Vet Med Surg 1: 43–50.
24. Madewell BR (1999). Diagnosis, assessment of prognosis, and treatment of dogs with lymphoma: The sentinel change (1973–1999). J Vet Intern Med 13: 393–394.
25. Modiano JF, Smith R 3rd, Wojcieszyn J, Thomas JS, Rosenbaum BA, Ball C, Nicholds EA, Anthony MA, Barton CL (1998). The use of cytochemistry, immunophenotyping, flow cytometry, and *in vitro* differentiation to determine the ontogeny of a canine monoblastic leukaemia. Vet Clin Pathol 27: 40–49.
26. Perez Alenza MD, Rutteman GR, Kuipers-Dijkshoorn NJ, Pena L, Montoya A, Misdorp W, Cornelisse CJ (1995). DNA flow cytometry of canine mammary tumours: the relationship of DNA ploidy and S-phase fraction to clinical and histological features. Res Vet Sci 58: 238–243.
27. Rothe G, Schmitz G, Adorf D, Barlage S, Gramatzki M, Hanenberg H, Höffkes HG,, Janossy G, Knüchel R, Ludwig WD, Nebe T, Nerl C, Orfao A, Serke S, Sonnen R, Tichelli A, Wörmann B (1996). Consenus protocol for the flow cytometric immunophenotyping of hematopoitic malignancies. Leukemia 10: 877–895.
28. Ruslander DA, Gebhard DH, Tompkins MB, Grindem CB, Page RL (1997). Immunophenotypic characterisation of canine lymphproliferative disorders. In Vivo 11: 169–172.
29. Rutteman GR, Cornelisse CJ, Dijkshoorn NJ, Poortman J, Misdorp W (1988). Flow cytometric analysis of DNA ploidy in canine mammary tumors. Cancer Res 48: 3411–3417.
30. Schwyn U, Crompton NEA, Blattmann H, Hauser B, Klink B, Parvis A, Ruslander D, Kaser-Hotz B (1998). Potential tumour doubling time: Determination of T_{pot} for various canine and feline tumours. Vet Res Commun 22: 233–247.
31. Teske E, Rutteman GR, Kuipers-Dijkshoorn NJ, van Dierendonck JH, van Heerde P, Cornelisse CJ (1993). DNA ploidy and cell kinetic characteristics in canine non-Hodgkin's lymphoma. Exp Hematol 21: 579–584.
32. Teske E, van Heerde P, Rutteman GR, Kurzman ID, Noore PF, MacEwen EG (1994). Prognostic factors of treatment of malignant lymphoma in dogs. Am J Vet Med Assoc 205: 1722–1728.
33. Teske E, Wisman P, Moore PF, van Heerde P (1994). Histologic classification and immunophenotyping of canine non-Hodgkin's lymphomas: Unexpected high frequency of T cell lymphomas with B cell morphology. Exp Hematol 22: 1179–1187.
34. Vail DM, Kisseberth WC, Obradovich JE, Moore FM, London CA, MacEwen EG, Ritter MA (1996). Assessment of potential doubling time (T_{pot}), argyrophilic nucleolar organizer regions (AgNOR), and proliferating cell nuclear antigen (PCNA) as predictors of therapy response in canine non-Hodgkin's lymphoma. Exp Hematol 24: 807–815.
35. Vernau W, Moore PF (1999). An immunophenotypic study of canine leukemias and preliminary assessment of clonality by polymerase chain reaction. Vet Immunol Immunopathol 69: 145–164.
36. Verschueren CP, Rutteman GR, Kuipers-Dijkshoorn

NJ, Sjollema BE, Vos JH, van Dijk JE, Cornelisse CJ (1991). Flow-cytometric DNA ploidy analysis in primary and metastatic canine thyroid carcinomas. Anticancer Res 11: 1755–1761.
37. Vindeslov LL, Christensen IJ, Nissen NI (1983). A detergent-trypsin method for the preparation of nuclei for flow cytometric DNA analysis. Cytometry 3: 323–327
38. Weiss DJ, Bauvelt M, Sykes J (2000). Flow cytometric evaluation of canine bone morrow differential cell counts. Vet Clin Pathol 29: 97–104.
39. Weiss DJ (2001). Flow cytometric and immunophenotypic evaluation of acute lymphocytic leukaemia. J Vet Intern Med 15: 589–594.
40. Weiss DJ (2001). Evaluation of proliferative disorders in canine bone marrow by use of flow cytometric scatter plots and monoclonal antibodies. Vet Pathol 38: 512–518.
41. Weiss DJ (2002). Application of flow cytometric techniques for veterinary clinical haematology. Vet Clin Pathol 31: 72–78.

Address for correspondence: Katja Culmsee, Clinic of Small Animals, School of Veterinary Medicine Hannover, Bischofsholer Damm 15, 30173 Hannover, Germany
Phone: +49 511 856 7251; Fax: +49 511 856 7686
E-mail: katja.Cculmsee@tiho-hannover.de

Methods in Cell Science **24**: 55–60 (2002).

Flow cytometric monitoring of drug resistance in human tumor cells

Awtar Krishan
Division of Experimental Therapeutics, University of Miami Medical School, Miami, FL, USA

Abstract. Recent studies have identified a family of glycoproteins which moulate cellular transport of antibiotics, alkaloids and drugs used in cancer chemotherapy. By facilitating efflux of drugs from the intracellular domain, these proteins reduce cytotoxicity and thus confer drug resistance. With the availability of antibodies raised against these phenotypic markers of drug resistance, immunohistochemistry and flow cytometry has been used to study their distribution and expression in normal and tumor cells. As some of the drugs used in cancer chemotherapy and other dyes which are substrates for this efflux pump are fluorescent, laser flow cytometry can be used for rapid quantitation of cellular retention, efflux and heterogeneity in drug transport of a tumor cell population. This method can also be used to screen drugs which may block efflux of a chemotherpeutic drug and thus increase chemosensitivity of a drug resistant tumor. In the present report flow cytometric methods for the study of drug transport and its modulation in tumor cells are discussed.

Key words: Drug resistance, Drug transport, Flow cytometry, Multiple drug resistance

1. Introduction

Recent studies [1–3, 7, 8, 20, 22, 23] have identified several glycoproteins of the ATP cassette super family that are involved in cellular transport of a variety of alkaloids (e.g, colchicine, vinblastine), antibiotics (e.g, doxorubicin, actinomycin D), hormones and carcinogens (e.g, benzypyrine). These glycoproteins are expressed in several normal cells (e.g, endothelial cells of the brain and testes) and in drug resistant tumor cells. A major role of these proteins is to control transport of hormones and efflux toxic compounds from the intracellular space and thus protect normal tissues from drug toxicity. In cancer chemotherapy, these proteins by effluxing therapeutic drugs can modulate cytotoxic response and thus render the cells resistant. Some of the well-known and studied drug resistance-modulating proteins expressed in drug resistant tumor cells are P-glycoprotein (Pgp, 8,20), Multidrug Resistance Associated Protein (MRP, 1,2), Lung Resistance Related Protein (LRP, 7,18,23) and the Breast Cancer Resistance Protein (BCRP, 3,22).

With the commercial availability of antibodies, immunohistochemistry became a major tool for detecting the expression of drug resistance proteins in histological sections; cell smears and in tissue digests. Subsequently flow cytometry was used for detection of their expression in leukemic cells and single cell suspensions of solid tumors. Several drugs, including antibiotics, alkaloids and dyes have been identified as substrates for the efflux pump. Some of these drugs will compete for the efflux pump and enhance retention of a chemotherapeutic drug and thus render resistant tumors sensitive [5, 9, 14, 24–26].

As several well-known fluorescent drugs are substrates for the drug resistance protein related efflux pump, we have developed flow cytometric procedures for monitoring their retention and efflux in tumor cells. Our initial studies were focused on cellular transport, retention and efflux of the fluorescent antitumor antibiotics, doxorubicin and daunomycin [12, 14] and other dyes [9, 17] in drug sensitive and resistant cell lines. We subsequently used these methods for monitoring of drug retention in tumor cells from patients [15, 16] as well as to study the effect of efflux blockers on their retention [10, 11, 13, 21]. The flow cytometric methods developed for monitoring of drug resistance and the effect of modulators which enhance drug retention are rapid, relatively inexpensive and can provide data by analysis of a large number of cells and differentiate between the expression and response of different subpopulations in a heterogeneous tumor population. In the present report we will illustrate the use of these flow cytometric methods for monitoring drug retention, efflux and modulation of drug resistance phenotype in tumor cells.

2. Material and methods

For flow cytometric analysis of drug retention and efflux in human tumor cells, it is useful to have available a cell line with known drug resistant phenotype, a fluorescent drug or dye which can be used as a substrate for the efflux pump and a second drug or agent which can act as an efflux blocker.

2.1. *Drug efflux substrates*

Several antibiotics used in cancer chemotherapy as well as some other drugs are fluorescent and as they are effluxed from drug resistant cells they can be used for the study of cellular transport and retention by flow cytometry [4, 6, 9, 13, 17, 19]. Fluorochromes commonly used for flow cytometric analysis using excitation from a 488 nM laser line include doxorubicin (adriamycin, NSC-123127, Adria Labs, Columbus, OH), daunomycin (Cerubidine, NSC-821151, Ives Labs, NY) and rhodamine 123 (Calbiochem). Doxorubicin and daunomycin bind to DNA by intercalation and most of their fluorescence is confined to the nucleus while rhodamine 123 is a mitochondrial-binding agent. At equimolar concentrations, rhodamine 123 has greater fluorescence than daunomycin, which in turn is brighter than doxorubicin. Several other drugs, which are effluxed in drug resistant cells, can be excited by UV illumination. The DNA binding dye, Hoechst 33342 (Calbiochem, Inc., San Diego, CA) and the calcium probe,Indo-AM are very good substrates for excitation with a high pressure mercury bulb or a UV laser line.

Stock solutions (1 mg/ml) of doxorubicin, daunomycin or rhodamine 123 are made in distilled water and stored frozen. Serial dilutions are made in phosphate buffered saline or Hanks' balanced salt solution to achieve the final concentration needed. Doxorubicin is normally used in final concentrations of 1–5 μg/ml, daunomycin at 0.1–2.0 μg/ml while rhodamine 123 can be used at much lower concentrations of 0.01–0.1 μg/ml. Cells are incubated with doxorubicin (1–3 h) or daunomycin or rhodamine for 10–30 min. before analysis.

2.2. *Drug efflux blockers*

Drug efflux blockers are chemicals that by blocking drug efflux enhance cellular retention and sensitivity [5, 9, 14, 24–26]. Some of the well-known efflux blockers include calcium channel active drugs (e.g., verapamil), phenothiazines, quinine, cyclosporins, dipyridamole, tamoxifen, as well as monoclonals against the p-glycoprotein. Most of the efflux blockers are used in the concentration range of 1–10 μg/ml.

2.3. *Indicator cell lines*

Although one can compare drug retention of cells incubated with the substrate drug alone and that of cells co-incubated with the substrate and the efflux blocker, it is advisable to have a well characterized pair of cell lines with known drug resistance available in the lab. These lines can be used for testing the reagents and in general can be included as a positive control when comparing drug retention and efflux of human tumor cells. Several well-characterized cell lines with the drug resistant phenotype have been reported in the literature [13] and we have extensively used suspension cultures of murine leukemic P388 and its doxorubicin-resistant cell line P388/R84 [21] for this purpose. These cell lines are readily available, easy to maintain in log-phase cultures and can be used for control purposes.

Single cell suspensions at the optimal concentration of 10^6/ml from log phase cultures are incubated with the fluorochromes and a parallel sample is co-incubated with the fluorochrome and the efflux blocker. The clinical sample to be analyzed for drug efflux which may consist of a single-cell suspension prepared from bone marrow aspirates, cells retrieved from ascites, or pleural fluid or cells disassociated after enzymatic digestion of solid tumors are incubated in a similar fashion. Cells are recovered by centrifugation and filtered through nylon mesh (40 micron) to remove clumps. The supernatant fluid is aspirated (after centrifugation) and the cell pellets resuspended and washed in Ca^{2+}- and Mg^{2+}-free phosphate-buffered saline. After centrifugation, the pelleted cells are resuspended in fresh tissue culture medium supplemented with 10% heat-inactivated fetal bovine serum. To remove erythrocytes from bone marrow aspirates and peripheral blood samples, specimens are diluted and layered over Histopaque-1077 (Sigma Diagnostics, St. Louis, MO). Mononuclear cells are recovered and washed before incubation with the drugs at 37 °C.

2.4. *Flow cytometric analysis*

Most of our earlier published work on flow cytometric monitoring of drug retention was carried out on a Becton and Dickinson FacScan or Coulter Electronics XL cytometer. Filter combination and protocols were used to generate scattergrams and list mode data on forward vs. side scatter and forward scatter vs. fluorescence. In some experiments we have used time as one of the parameters to generate plots of time vs. cellular fluorescence.

Samples are analyzed at the flow rate of 100–200 cells/second and a typical analysis is based on examination of 10,000 cells. Drug fluorescence is measured on a log scale while the scatter signals are on a linear scale. Filter and photo multiplier combinations used are determined by the characteristics of the fluorochromes used as the substrate for the efflux pump.

An aliquot of drug sensitive cells with the optimal concentration and time of incubation is used to collect the peak of the log fluorescence between the second and the third log on the X-axis. The drug resistant cells with or without the efflux blocker are then analyzed to see if the efflux blocker is working and the lower fluorescence of the resistant cells is greater than that of the background noise. In most of the drug sensitive and resistant-paired cell lines, fluorescence peak of the resistant cells without the blocker is registered in middle of the second log with the cells co-incubated with an efflux blocker recording log higher peak fluorescence.

It should be noted that drug resistant cells with damaged cell membrane would record higher fluorescence than that of the cells with functional efflux pump. Thus as shown in one of our methodology papers [10] it is important to discriminate between the membrane damaged resistant cells with bright fluorescence and drug sensitive cells which do not efflux the drug. For this purpose one can use the forward scatter vs. fluorescence as a gate to exclude the dead cells. A better method based on the use of isotonic propidium iodide for identification of the membrane damaged cells has been described in our earlier publication [10] and shown in Figure 2).

3. Results and discussion

Following three figures are used to illustrate the type of data one can accumulate by flow cytometric analysis of drug uptake and retention in drug sensitive and resistant cells.

Contour plots in Figure 1 compare cellular fluorescence of drug sensitive P388 and drug resistant P388/R84 cells incubated with 0.1 μg/ml of rhodamine-123. On the x-axis time is recorded in seconds while on the Y-axis cellular fluorescence is shown on a 4-decade log scale. The rapid uptake of the rhodamine is seen in the P388 cells while in the resistant cells cellular retention of rhodamine fluorescence is lesser presumably due to the presence of the P-gp drug efflux pump. After 10 min of incubation, the amount of rhodamine fluorescence in P388 cells is nearly a log higher than that of the P388/R84 cells.

In scatter plots and histograms of Figure 2, the effect of cells with damaged cell membrane on flow cytometric analysis of daunorubicin fluorescence is shown. The scatter plots in A and E show drug fluorescence vs. forward scatter. Heterogeneity of drug retention is evident in both of the drug sensitive and the resistant P388 cells. In histograms B and F, the P388/R84 cells show the presence of two distinct populations while the P388 cells seem to have only a single population. In cells co-incubated with daunorubicin and isotonic propidium iodide (shown in Figure 2 C, D, G, H), the bright propidium iodide fluorescence of the cells with damaged and leaky cell membrane makes it easy to discriminate between these cells and those with low drug retention due to the presence of the efflux pump (panel G and H) and cells with high drug retention but intact membranes (panel C and D).

In Figure 3 contour plots of doxorubicin fluorescence in cells collected from pleural fluid of two lung cancer patients are compared. Cells were incubated with doxorubicin (panels A, D, G) or in the presence of 10 μg/ml of chlorpromazine (B, E, H) or verapamil (panels C, F, I). Note in the top panel incubation with chlorpromazine did enhance drug retention (Compare A with B in top figure) while verapamil had no effect on drug efflux (panel A vs. C, top figure). In contrast to the cells from the patient shown in the top panel, cells from the second patient (shown in the bottom

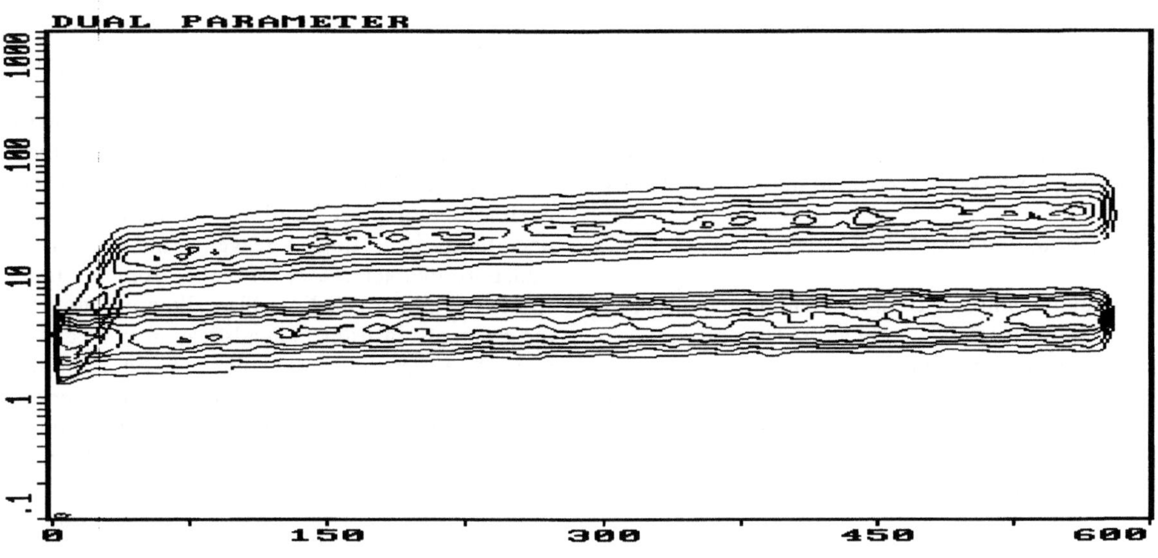

Figure 1. Shows uptake of rhodamine in P388 drug sensitive and resistant cells. On the X-axis time is recorded in seconds while on the Y-axis, cellular fluorescence of rhodamine 123 is shown. The top curve is of the drug sensitive cells while the bottom is of the drug resistant P388/R84 cells (From Krishan,10).

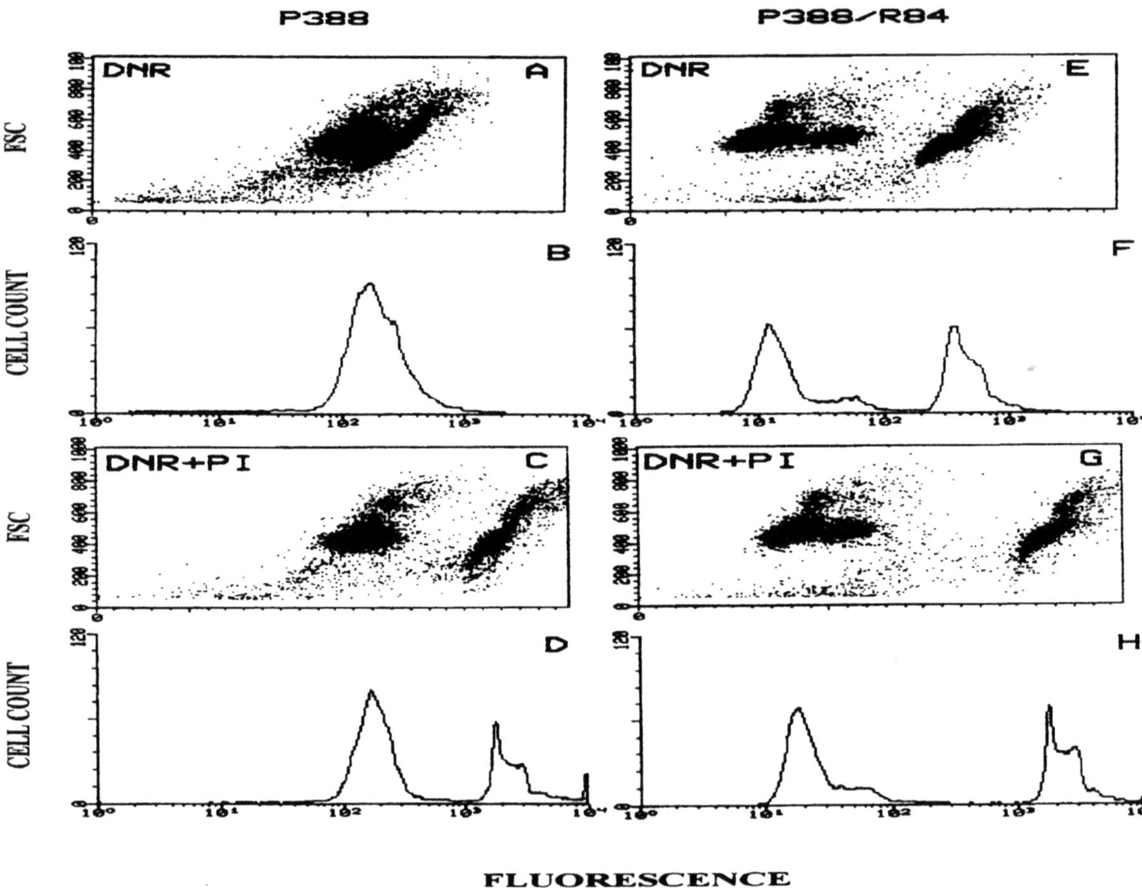

Figure 2. Shows the effect of membrane permeable cells on monitoring of drug retention and efflux in drug sensitive and resistant cells. Propidium iodide is used to identify dead cells (dye permeable) and distinguish them from other cells with drug efflux (From Krishan, 10).

panel) were relatively resistant to the efflux blocking effect of chlorpromazine or verapamil. The two panels in the middle and on the right side show data on drug retention and modulation in cells from serial samples tested approximately within one month of the initial sample (shown in the panel on the left). Emergence of drug resistant cells, which do not show drug fluorescence either on incubation with doxorubicin alone or on co incubation with the efflux blockers, is suggested by data in these contour plots.

3.1. *Precautions and pitfalls*

Although monitoring of drug retention and efflux by flow cytometry is relatively easy, data collected can be erroneous if precautions are not taken in setting up the conditions for the studies. In an earlier review article we have discussed precautions and guidelines, which should be followed for proper preparation of samples for flow analysis [10]. Here we will briefly summarize some of these factors. In dealing with clinical material such as tumor biopsies or fluids, one of the major problems is 'unpredictability' as to what time the sample will arrive and under what condition. As drug transport studies are carried out on live cells, we have occasionally received samples for such analysis already fixed in formalin. Similarly a sample may be necrotic or coagulated (e.g., bone marrow or ascites) or hardly sufficient for collecting any meaningful data. Thus, it is important to educate the clinical colleagues as to the requirements for proper collection, handling and transport of the specimens.

In the lab, it is essential to have proper controls (e.g, indicator cell lines) that can be used to validate data on the clinical sample. As discussed earlier availability of paired cell lines removes a lot of uncertainty in testing of reagents and determining transport characteristics of cells from a clinical specimen. These cell lines also allow one to set up the protocols, instrument amplification and gains and make sure that the signal collected is not confused with data from auto fluorescence and fluorescence of membrane permeable and dead cells. These cell lines are also important for testing the drugs used for transport and efflux blocking studies.

Data from functional assays may be compromised by heterogeneity in drug retention and response to efflux blockers of the different sub-populations and the need to distinguish between the resistant cell with permeable presumably damaged cell membrane and

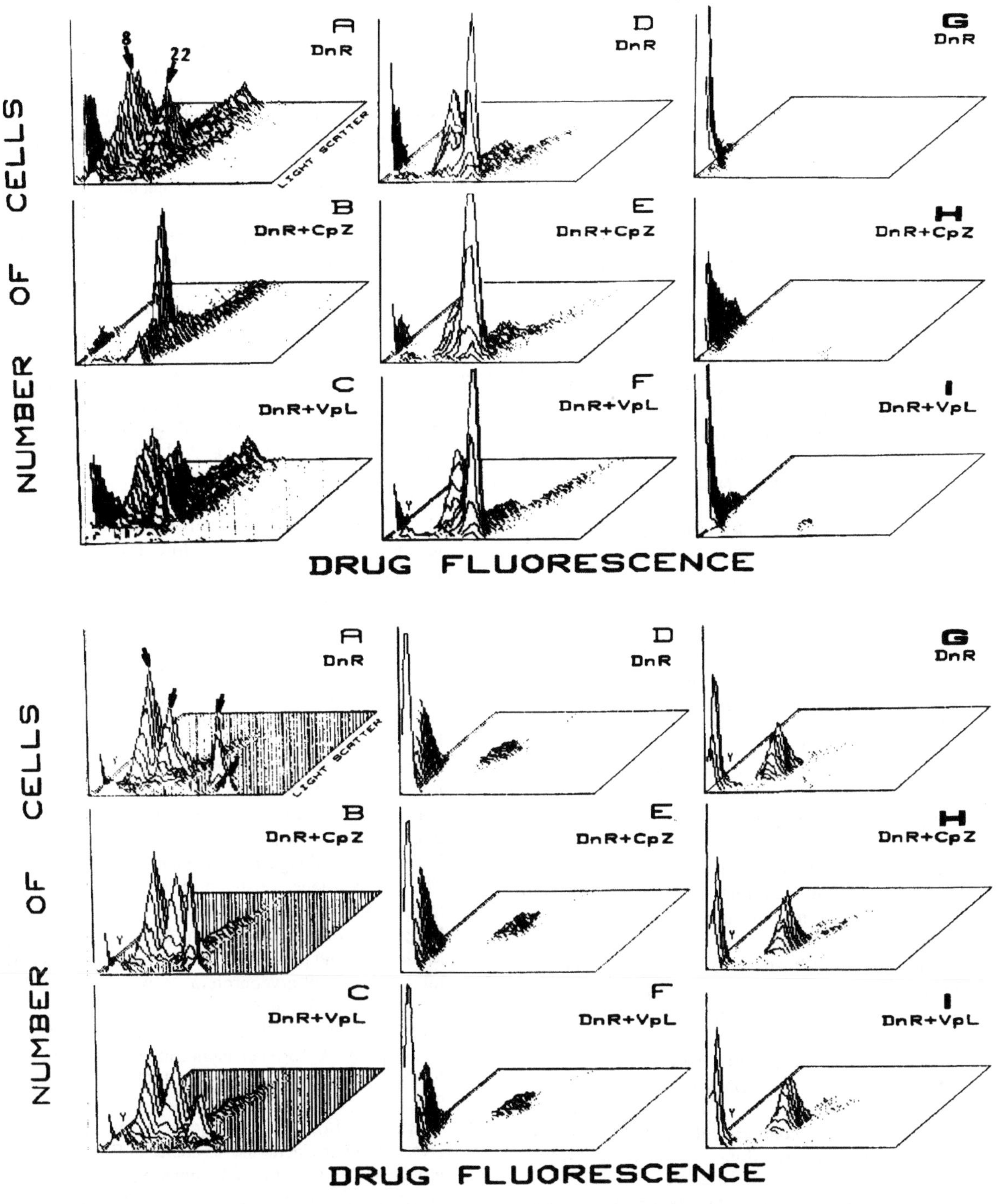

Figure 3 compares drug uptake and efflux blocking in cells from two lung cancer patients (From Krishan et al. 15).

high drug fluorescence and the live cells with high drug retention.

Although it would be ideal to study drug transport and resistance of the proliferating and clonogenic tumor cells, it is often difficult to distinguish between these populations and the infiltrating cells such as lymphocytes and macrophages. The use of antibodies to identify the stromal cells has been attempted in several published studies and shown in one of our earlier publications [16].

Acknowledgements

This work was supported by a Department of Army grant #DAMD17-00-1-0342.

Notes on suppliers

Most of the substrates, antibiotics, dyes and efflux blockers used in our flow cytometric studies are available from Sigma Chemicals, St Louis, MO and from Molecular

Probes, Inc, Eugene, OR. We had purchased the following drugs from the specific sources listed here: Doxorubicin (adriamycin, NSC-123127, Adria Labs, Columbus, OH). Daunomycin (Cerubidine, NSC-821151, Ives Labs, NY). Rhodamine 123 (Calbiochem, San Diego, CA). Hoechst 33342 (Calbiochem, Inc., San Diego, CA).

References

1. Almquist KC, Loe DW, Hipfner DR et al. (1995). Characterization of the Mr 190,000 multidrug resistance protein (Mrp) in drug-selected and transfected human tumor cells. Cancer Res 55: 102–110.
2. Deeley RG, Cole SP (1997). Function, evolution and structure of multidrug resistance protein (MRP). Semin Cancer Biol 8: 193–204.
3. Doyle LA, Yang W, Abruzzo LV et al. (1998). A multidrug resistance transporter from human MCF-7 breast cancer cells. Proc Natl Acad Sci USA 95: 15665–15670.
4. Frey T, Yue S, Haugland RP (1995). Dyes providing increased sensitivity in flow-cytometric dye-efflux assays for multidrug resistance. Cytometry 20: 218–227.
5. Ganapathi R, Grabowski D, Rouse W et al. (1981). Differential effect of the calmodulin inhibitor trifluoperazine on cellular accumulation, retention, and cytotoxicity of anthracyclines in doxorubicin (adriamycin)-resistant P388 mouse leukemia cells. Cancer Res 44: 5056–5061.
6. Hollo Z, Homolya L, Davis CW et al. (1994). Calcein accumulation as a fluorometric functional assay of the multidrug transporter. Biochim Biophys Acta 1191: 384–388.
7. Izquierdo M, Scheffer G, Flens M et al. (1996). Broad distribution of the multidrug resistance-related vault lung protein in normal human tissues and tumors. Am J Pathol 148: 877–887.
8. Juliano RL, Ling V (1976). A surface glycoprotein modulating drug permeability in Chinese hamster ovary cell mutants. Biochim Biophys Acta 455: 152–162.
9. Krishan A (1987). Effect of drug efflux blockers on vital staining of cellular DNA with Hoechst 33341. Cytometry 8: 642–645.
10. Krishan A (2001). Monitoring of cellular resistance to cancer chemotherapy: drug retention and efflux. Methods Cell Biology 64: 193–209.
11. Krishan A, Fitz MC, Andritsch I (1997). Drug retention, efflux and resistance in tumor cells. Cytometry 29: 279–285.
12. Krishan A, Ganapathi R (1980). Laser flow cytometric studies on the intracellular fluorescence of anthracyclines. Cancer Res 40: 3895–3900.
13. Krishan A, Sauerteig A, Andritsch I, and Wellham L (1997). Flow cytometric analysis of the multiple drug resistance phenotype. Leukemia 11: 1138–1146.
14. Krishan A, Sauerteig A, Wellham L (1985). Flow cytometric studies on modulation of cellular adriamycin retention by phenothiazines. Cancer Res 45: 1046–1051.
15. Krishan A, Sridhar KS, Davila E et al. (1987). Patterns of anthracycline retention modulation in human tumor cells. Cytometry 8: 306–314.
16. Kunikane H, Zalupski MM, Ramachandran C et al. (1997). Flow cytometric analysis of P-glycoprotein expression and drug efflux in human soft tissue and bone sarcomas. Cytometry 30: 197–203.
17. Lampidis T, Munck J, Krishan A et al. (1985). Reversal of resistance to rhodamine 123 in adriamycin-resistant Friend leukemia cells. Cancer Res 45: 2626–2631.
18. Leith CP, Kopecky KJ, Chen IM (1999). Frequency and clinical significance of the expression of multi drug resistance proteins MDR1/P-gp, MRP1, and LRP in acute myeloid leukemia. A Southwest Oncology Group study. Blood 94: 1086–1099.
19. Lelong EH, Guzikowski AP, Haugland RP et al. (1991). Fluorescent verapamil derivative for monitoring activity of the multidrug transporter. Mol Pharmacol 40: 490–494.
20. Ling V (1992). P-glycoprotein and resistance to anticancer drugs. Cancer 69: 2603–2609.
21. Nair S, Singh SV, Samy TSA et al. (1990). Anthracycline resistance in murine leukemic P388 cells: Role of drug efflux and glutathione related enzymes. Biochemical Pharmacol 39: 723–728.
22. Ross DD, Gao Y, Yang W et al. (1997). The 95-kilodalton membrane glycoprotein overexpressed in novel multidrug resistant breast cancer cells is NCA, the nonspecific cross-reacting antigen of carcinoembryonic antigen. Cancer Res 57: 5460–5464.
23. Scheffer GL, Wijngaard PLJ, Flens MJ et al. (1995). The drug resistance related protein LRP is the human major vault protein. Nat Med 1: 578–582.
24. Sikic BI, Fisher GA, Lum BL et al. (1997). Modulation and prevention of multidrug resistance by inhibitors of P-glycoprotein. Cancer Chemother Pharmacol 40: Suppl S13–S19.
25. Slater LM, Sweet P, Stupecky M et al. (1986). Cyclosporin A corrects daunorubicin resistance in Ehrlich ascites carcinoma. Br J Cancer 54: 235–238.
26. Tsuruo T, Iida H, Tsukagoshi S et al. (1981). Overcoming of vincristine resistance in P388 leukemia *in vivo* and *in vitro* through enhanced cytotoxicity of vincristine and vinblastine by verapamil. Cancer Res 41: 1967–1972.

Author for correspondence: Awtar Krishan, Experimental Therapeutics Division (R-71), University of Miami Medical School, P.O. Box 01690, Miami, FL 33101, USA
Phone: 305-243-6553; Fax: 305-243-5555
E-mail: akrishan@med.miami.edu

Methods in Cell Science **24**: 61–64 (2002).

Androgen receptor expression in archival human breast tumors

Poonam Arya*, Ilia H. Andritsch & Awtar Krishan
Division of Experimental Therapeutics, University of Miami Medical School, Miami, FL, USA

Abstract. Hormone receptors play a major role in growth and hormonal therapy of breast and prostate tumors. Quantitative results from the ligand binding assays cannot determine heterogeneity in receptor expression nor can they discriminate between expression of the stromal and the tumor cells. Availability of antibodies to hormone receptors has led to the development of immunohistochemistry as a standard method for monitoring of hormone receptor expression under a microscope. However, this method is based on examination of a small number of cells. Laser flow cytometry has been extensively used for monitoring of receptor expression in human liquid tumors. As most of the hormone receptor expression is nuclear, we have developed methods for flow cytometric analysis of receptor expression in nuclei isolated from enzyme treated paraffin sections. The present report based on gated analysis of androgen receptor expression in nuclei isolated from archival formalin fixed/paraffin embedded breast tumors shows that receptor expression in aneuploid sub-populations is greater than that of the diploid cells.

Key words: Androgen receptor, Breast tumor, Flow cytometry

Abbreviations: MCF = Mean channel log fluorescence

1. Introduction

Human breast and prostate cells have estrogen and androgen receptors, which react to the presence of circulating hormones to modulate cell growth and function [1–3, 7, 9–11]. Hormone therapy is used for clinical management and prevention of breast and prostate cancer [2, 3, 8–10, 12].

Determination of hormone receptor level in a tumor biopsy is of both diagnostic and therapeutic value for the management of these hormone sensitive tumors. In the past ligand-binding assays were used to determine the level of hormone receptor expression in breast and prostate tumors [4]. However, with the availability of antibodies raised against the receptor proteins, immunohistochemistry has replaced the ligand-binding assays. Although data from ligand-binding assays is quantitative, this procedure does not allow for discrimination of receptor expression between normal stromal and tumor cells. In contrast the immunohistochemical procedures in general are not quantitative and analysis is based on examination of a small number of cells on a microscope slide [13, 14].

Recently, flow cytometric methods have been developed for the quantitative determination of estrogen, progesterone and androgen receptor expression in paraffin embedded human breast and prostate tumors [7, 11]. In contrast to immunohistochemistry, flow cytometric analysis can provide quantitative data on antigen expression and correlate it with other cellular parameters such as DNA aneuploidy or expression of other hormone receptors [7, 11]. We have recently developed methods for analysis of androgen receptor expression in human prostate tumors [7] and reported on correlation of receptor expression with prostate tumor grade and stage [1]. Several recent studies have used immunohistochemical methods to study androgen receptor expression in breast tumors [5, 6]. In the present study we have used flow cytometric methods for the analysis of androgen receptor expression in archival human breast tumors.

2. Material and methods

Thin (10 μM) and thick (50 μM) sections were cut from formalin fixed-paraffin embedded breast tumor blocks. A hematoxylin/eosin stained slide was used to confirm the presence of tumor cells in sections used for flow cytometric analysis.

2.1. *Antigen retrieval*

Thick sections deparaffinized in xylene and rehydrated in a descending ethanol series were resuspended in citrate buffer (0.01 M, pH 6.0) and heated for 120 min at 80 °C in a water bath. After 30 min of cooling at room temperature, the sections were washed in phosphate buffered saline (PBS), collected by centrifugation and processed for isolation of nuclei by enzyme digestion.

2.2. *Tissue digestion*

Tissue digestion was performed in 0.05% pepsin (Sigma, Cat. No 7012, pH 1.65) in a water bath at 37 °C for 30 min. Samples were vortexed every 5

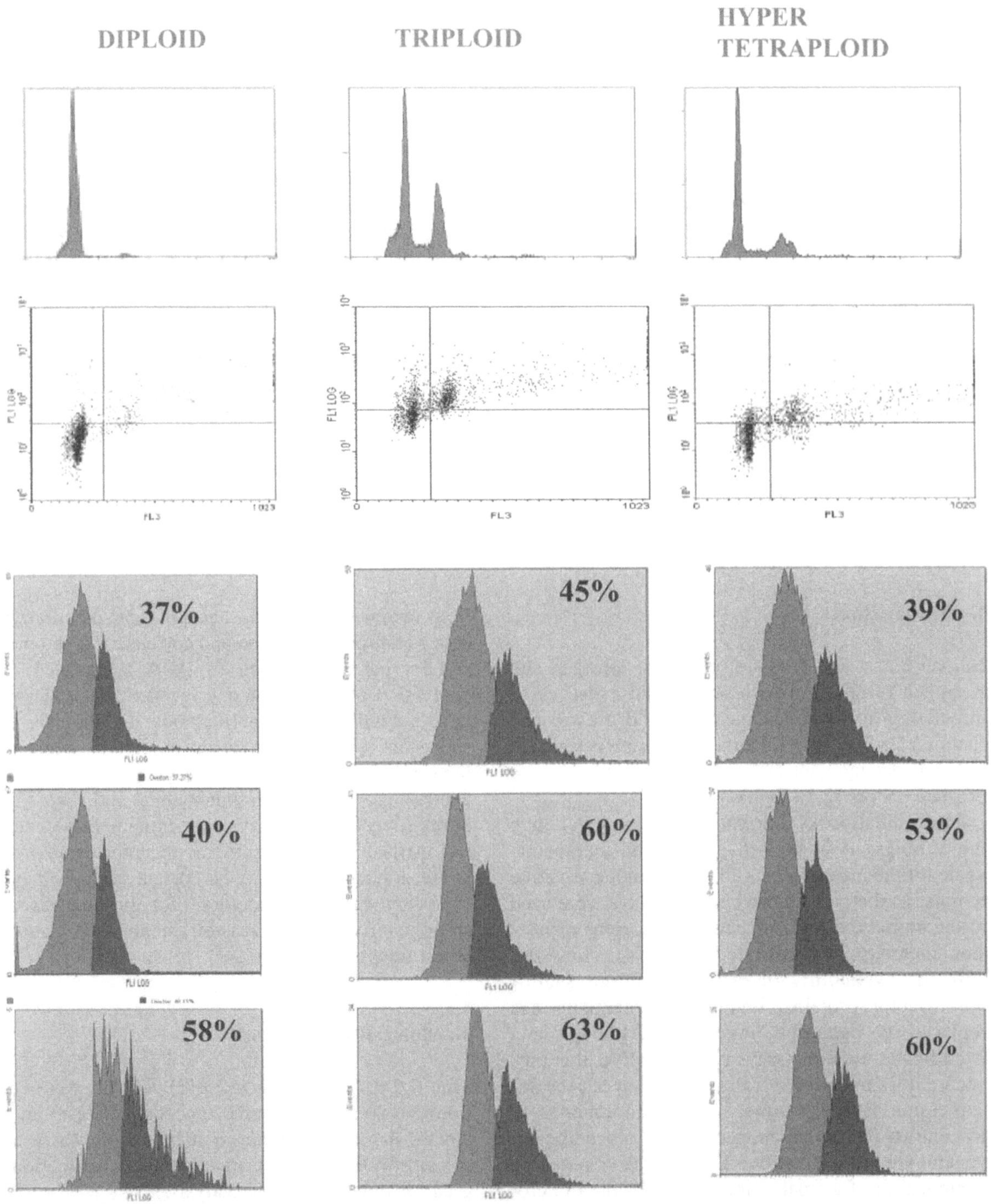

Figure 1. DNA distribution histograms of the three representative tumors show the presence of triploid and hyper-tetraploid sub-populations in two tumors (middle and right panel). Scattergrams in the second row are of DNA content vs. androgen expression. The horizontal line is the gate used to subtract 95% of fluorescent cells from the isotype controls. The overlay histograms in the bottom three rows are of androgen expression the isotype controls (light shade) and the mAB (darker shade) treated cells. The top set of overlay histograms is of the total population while the middle and the bottom row is of the gated triploid and hyper-tetraploid sub-populations.

min. during digestion. Proteolytic action was terminated by the addition of 3% fetal bovine serum (FBS) in PBS. The resulting digest was filtered through a 40-μm-nylon mesh, centrifuged and washed in PBS.

2.3. *Staining for androgen receptor expression and DNA analysis*

The isolated nuclei were incubated for 18h with 150 μl of anti-AR antibody (MU256-UC, Clone F39.4.1, Biogenex, San Ramon, CA) at 1:45 dilution in PBS at 37 °C. The negative isotype control used for the anti-AR antibody was normal mouse IgG1 (MOPC 21, Sigma Cat No. M5284) at 1:180 dilution in PBS. Samples were washed with PBS containing 0.05% Triton-X and 3% goat serum. The resuspended nuclear pellet was stained with 150 μl of FITC conjugated goat anti-mouse IgG antibody (1:80 dilution in PBS, Cat No. 4143, Sigma Chemicals Co.) for 45 min at room temperature in dark. After centrifugation, nuclei were washed with PBS (0.05% Triton-X) and resuspended in PBS for flow cytometric analysis. For simultaneous monitoring of AR expression and nuclear DNA content, AR-FITC stained nuclei were incubated with propidium iodide (25 μg/ml + 0.5 mg/ml RNAse in PBS) for 30 min.

2.4. *Flow cytometry*

Samples were analyzed on a XL-*MCL* Flow cytometer (Beckman/Coulter Electronics, USA) with the standard argon ion laser excitation and filter selection for FITC/propidium iodide combination. The relative fluorescence intensity (MCF) was determined by dividing the mean log channel fluorescence value of the antibody-reacted samples by that of the isotype controls. To determine the percent of AR positive nuclei, the subtract function (by Overton's method) in EXPO-32 software (Beckman/Coulter, USA) was used.

3. Results and discussion

Figure 1 is based on the analysis of nuclei isolated from three different breast tumor samples for DNA content and androgen receptor expression. DNA distribution histograms in the first row show that in contrast to the diploid tumor on the left side, the two other tumors in middle and right had triploid and hyper-tetraploid subpopulations.

Scattergrams in the second row show DNA content (x-axis) vs. androgen receptor expression of the nuclei from these tumors. The horizontal line indicates the electronic gate set to exclude 95% of fluorescent cells from the isotype controls. The vertical line is drawn between the diploid and the aneuploid cells.

The overlay histograms in the bottom three rows of this figure are of the isotype (light color) and the mAB (darker shade) treated samples. The first row shows the percent of androgen positive cells in the total sample. In the second row the percentage of androgen positive cells in the diploid population is shown while the bottom row shows the percent of positive cells in the aneuploid sub-population.

Analysis of more than 35 breast tumors and their aneuploid sub-populations, have shown that in general aneuploid tumor cells in breast tumor have higher percentage of androgen positive cells than the diploid populations. Our data suggests that as the tumors become aneuploid, the percentage of androgen positive cells also increases.

3.1. *Precautions*

Antigen retrieval and enzyme digestion if not carried out properly can damage DNA and the receptors leading to the generation of artifacts. Thus the temperature of the water bath and the length of treatment for antigen retrieval are critical elements and should be closely monitored. Aggressive enzyme treatment can lead to proteolytic digestion of the receptors and one has to balance the need for preserving the receptor expression with need to isolate as many intact nuclei as possible from the paraffin embedded sections.

Notes on suppliers

Pepsin (Sigma, Cat No 7012). Anti-AR antibody (MU256-UC, Clone F39.4.1, Biogenex, San Ramon, CA), Mouse IgG1 (MOPC 21, Sigma Cat No. M5284), FITC conjugated goat anti-mouse IgG antibody (1:80 dilution in PBS, Cat No. 4143, Sigma Chemicals Co.), XL-*MCL* Flow cytometer (Beckman/Coulter Electronics, USA), EXPO-32 software (Beckman/Coulter, USA).

References

1. Adiga SK, Andritsch I, Rao RV, Krishan A (2002). Androgen receptor expression and DNA content of paraffin embedded archival human prostate tumors. Clinical Cytometry 50: 25–30.
2. Allred DC, Bustamante MA, Daniel CO, Gaskill HV, Cruz AB Jr. (1990). Immunocytochemical analysis of estrogen receptors in human breast carcinomas. Evaluation of 130 cases and a review of the literature regarding concordance with the biochemical assay and clinical relevance. Arch Surg 125: 107–134.
3. Benson RCJ, Gorman PA, O'Brien PC, Holicky EL, Veneziale CM (1987). Relationship between androgen receptor binding activity in human prostate cancer and clinical response to endocrine therapy. Cancer (Phila.) 59: 1599–1606.
4. Brolin J, Skoog L, Ekman P (1992). Immunohistochemistry and biochemistry in detection of androgen, progesterone, and estrogen receptors in benign and

malignant human prostatic tissue. Prostate 20: 281–295.
5. Gatalica Z (1997). Immunohistochemical analysis of apocrine breast lesions. Consistent over-expression of androgen receptor accompanied by the loss of estrogen and progesterone receptors in apocrine metaplasia and apocrine carcinoma *in situ*. Pathol Res Pract 193(11–12): 753–758
6. Isola JJ (1993). Immunohistochemical demonstration of androgen receptor in breast cancer and its relationship to other prognostic factors. J Pathol 170(1): 31–35.
7. Krishan A, Oppenheimer A, You W, Dubbin R, Sharma D, Lokeshwar BL (2000). Flow cytometric analysis of androgen receptor expression in human prostate tumors and benign tissues. Clin Cancer Res 6: 1922–1930.
8. Long, RJ, Roberts KP, Wilson MJ, Ercole CJ, Pryor JL (1997). Prostate cancer: a clinical and basic science review. J Androl 18: 15–20.
9. Pichon MF, Broet P, Magdelenat J, Delarue JC, Spyratos F, Basuyau JP, Saez S, Rallet A, Courriere P, Millon R, Asselain B (1996). Prognostic value of steroid receptors after long-term follow-up of 2257 operable breast cancers. Br J Cancer 73: 1545–1551.
10. Prins GS, Sklarew RJ, Pertschuk LP (1998). Image analysis of androgen receptor immunostaining in prostate cancer accurately predicts response to hormonal therapy. J Urol 159: 641–649.
11. Redkar AA, Krishan A (1999). Flow cytometric analysis of estrogen, progesterone receptor expression and DNA content in formalin-fixed, paraffin-embedded human breast tumors. Cytometry 38: 61–69.
12. Sadi MV, Barrack ER (1991). Androgen receptor and growth fraction in metastatic prostatic cancer as predictors of time to tumor progression after hormonal therapy. Cancer Surv 11: 195–215.
13. Sweat SD, Pacelli A, Bergstralh EJ, Slezak JM, Bostwick DG (1999). Androgen receptor expression in prostatic intraepithelial neoplasia and cancer. J Urol 161: 1229–1232.
14. Tilley WD, Lim-Tio SS, Horsfall DJ, Aspinall JO, Marshall VR, Skinner JM (1994). Detection of discrete androgen receptor epitopes in prostate cancer by immunostaining: measurement by color video image analysis. Cancer Res 54: 4096–4102.

Address for correspondence: Poonam Arya, Experimental Therapeutics Division (R-71), University of Miami, School of Medicine, P.O. Box. 01690, Miami, FL 33101, USA
Phone: 305-243-6553; Fax: 305-243-5555
E-mail: poonam_arya@hotmail.com)

Methods in Cell Science **24**: 65–68 (2002).

Membrane oxidative damage and apoptosis in cervical carcinoma cells of patients after radiation therapy

S. M. Bhosle[1], B. N. Pandey[2], N. G. Huilgol[1] & K. P. Mishra[2]
[1] *Department of Radiation Oncology, Dr. Balabhai Nanavati Hospital, Mumbai 400 056, India;*
[2] *Radiation Biology and Health Sciences Division, Bhabha Atomic Research Centre, Mumbai 400 085, India*

Abstract. This article describes evaluation of plasma membrane fluidity and intracellular SOD with relation to apoptotic death of cervical carcinoma cells after radiation therapy. Cells from biopsies of cancer patients (stage IIIB) prior to and 24 h after radiation dose of 2 Gy were examined. Plasma membrane fluidity, measured by fluorescence polarization of DPH incorporated into lipid bilayer and superoxide dismutase (SOD) activity, determined by epinephrine method, showed significant decrease but per cent apoptotic cells, as determined by annexin-V and TUNEL methods, were found increased by two folds after radiotherapy. It is suggested that decrease in DPH polarization in membrane, reduction in SOD activity and increased apoptosis in cervical cells of cancer patients treated with radiation may be consequent to oxidative damage induced by reactive oxygen species (ROS), which may have implications in developing predictive protocol in cancer radiotherapy.

Key words: Apoptosis, Bilayer fluidity, Cervical carcinoma, Radiotherapy, Reactive oxygen species

Abbreviations: DMEM = Dulbecco's Modified Eagle Medium; DMSO = dimethyl sulphoxide; DPH = 1,6-diphenyl 1,3,5-hexatriene; ROS = Reactive oxygen species; SOD = superoxide dismutase; TUNEL = TdT *dUTP* nucleotide end labeling

1. Introduction

In recent years cellular apoptosis has been regarded as an important indicator of radiosensitivity, which has important implications in radiobiology and radiation oncology. For example, evaluation of relationship between cellular sensitivity and induction of apoptosis in tumor may provide a useful and effective guideline for designing appropriate protocol for radiotherapy. We have investigated membrane alterations, antioxidant enzyme level and induction of apoptosis in stage III uterine cervical cancer patients before and 24 h after radiation therapy. Apoptosis is characterized by structural and molecular features like membrane blebbing, externalization of phosphatidylserine (PS), nuclear shrinkage and DNA ladder formation [1]. Over the years a variety of assays have been developed which provide qualitative identification of apoptotic cells such as gel electrophoresis and dye exclusion methods. However, these methods lack specificity and are time consuming. To overcome these drawbacks, a promising flow cytometric method based on annexin V binding on surface of apoptotic cells has been developed which, together with DNA specific fluorescent probes, allows discrimination between apoptotic and necrotic cells [1, 2]. In addition, intensive research is directed to exploring the existence of possible interrelationship between changes in intracellular or membrane parameters such as lipid peroxidation, antioxidant enzymes, ROS etc. and the occurrence of apoptosis. It is hoped that susceptibility to oxidative damage to tumor may follow sensitivity to apoptotic course of death suggesting molecular alterations as a prognostic indicator for radiotherapy in clinic. Ionizing radiation e.g. γ-rays is known to cause oxidative damage to vital cellular components, namely, DNA, proteins and lipid molecules through production of ROS [3–4]. The understanding of radiation damage to cellular membrane with associated apoptotic cell death has considerable significance in cellular radiosensitivity and cancer radiotherapy. Fluorescence and ESR techniques have been extensively employed to study structural and dynamic changes in lipid bilayer induced by radiation [5–7] involving ROS and a relationship between γ-radiation induced membrane fluidity and apoptotic death in thymocytes has been indicated [8]. Present results suggest that significant alterations occurred in plasma membrane fluidity, SOD level and induced apoptotic cell death in cervical tumor cells of cancer patients after radiotherapy, which may form vital basis for long term goal of evaluating the usefulness of these parameters to determine whether prognosis of patients with stage III cervical cancer treated with radiation would be predictable.

2. Materials

Biopsy samples were collected from cervical cancer patients (stage IIIB) from Radiation Oncology Department of Nanavati Hospital, Mumbai. DPH (D4380) and epinephrine (E4642) were purchased from Sigma[1]. DMEM (12100-046) was obtained from GIBCO[2] and Annexin-V-FLUOS staining (12100-046) and TUNEL kit (1 684 7950) were purchased from Roche[3].

3. Procedures

Cell suspension was prepared under sterile condition from biopsy samples obtained from patients in DMEM by mincing with curved scissor and filtering through nylon filter. The viability of samples was measured by trypan blue (0.04%) dye exclusion method. Cells were washed and suspended in PBS, and further experiments were performed with the prepared suspension. For measurement of SOD activity, cells suspension in PBS (4×10^6 cells) was mixed with homogenizing buffer (0.25 M sucrose, 50 mM Tris-HCl, 50 μg/ml polymethyl sulphonyl fluoride, 10 μg/ml leupeptine, 10 μg/ml aproteinase, 10 μg/ml pepstatin, 5 mM EDTA, 0.3% w/w β-mercaptoethanol) followed by 15 min. sonication at 4 °C (cycle of 10 sec sonication and cooling). The protein concentration in samples was determined by Folin-Lowry's method [9]. SOD activity was determined in 20 μl lysate (2.9 μg protein) of control and radiation treated samples by epinephrine auto-oxidation method [10]. For fluorescence polarization studies, 1 μl of DPH (100 μM in DMSO) was incubated with 2.5 ml of cell suspension (1×10^6 cells) in PBS for 20 min at 25 °C followed by measurement of anisotropy as mentioned in Ref. [6]. Percentage apoptosis in cervical cells was determined by annexin-V and TUNEL method following the protocol provided with kits using fluorescence microscope (Optophot 2, Nikon, Japan). At least 400 cells were randomly counted from population to determine the percentage apoptosis.

4. Results and discussion

Radiation damaging effects on cells are generally believed to be mediated by free radical induced oxidative reactions with biomolecules but clear understanding of mechanism at molecular and cellular level remains elusive. The focus of our investigation has been to determine oxidative damage in plasma membrane and consequent relation with apoptotic response of cervical cancer cells from patients after radiotherapy. The oxidative damage of cells is often manifested in their plasma membrane alterations [4, 8, 11–12]. Tumor cell membrane changes were studied after labeling with lipid bilayer specific fluorescence probe, DPH and results showed an increase in probe anisotropy implying decreased bilayer fluidity after radiation treatment (Figure 1). The observed decrease in membrane fluidity after radiation therapy suggests involvement of radiation generated free radical reactions with membrane components leading to oxidative damage. A similar mechanism has been suggested by the authors to explain the γ radiation effects on phospholipid model membranes [6–7]. Living cells have evolved a variety of intracellular antioxidant enzymes for their defense [13] and SOD is predominantly known to neutralize induced superoxide radicals producing hydrogen peroxide by dismutation. Measurement of SOD in control and radiation treated cervical cells showed time dependent increase in absorbance peaking after 5 min. Moreover, in irradiated samples, increase in absorbance was found indicating decreased enzyme activity (Figure 2). Biopsy samples after pre- and post-irradiation from patients showed significantly decreased SOD activity 24 h after radiation therapy (Figure 3) suggesting a reduced level of enzymes following patient irradiation. Determination of apoptotic cervical cells after radiotherapy, as measured by annexin-V and TUNEL assays, showed two fold increased apoptosis in cells from radiation treated patients than in control (Table 1) indicating radiation initiated events leading to apoptotic cell death. Moreover, it needs to be noted that magnitude of apoptosis determined by annexin-

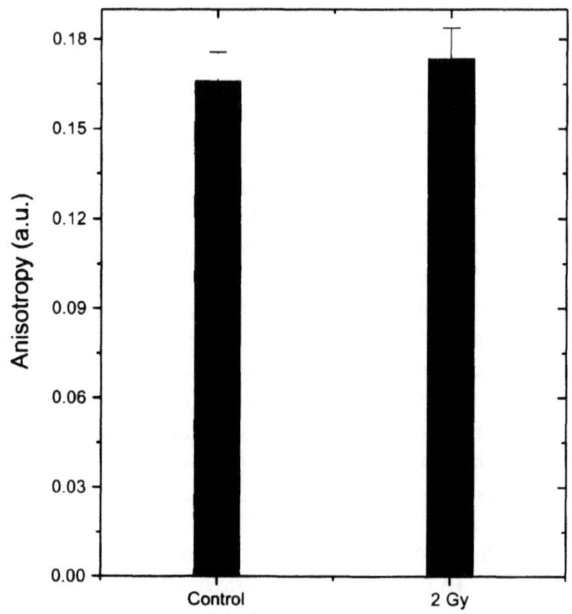

Figure 1. Membrane fluidity in radiation treated cervical tumor cells. Cell suspensions from biopsy samples prepared in PBS (1×10^6 cells) were labeled with DPH followed by measurement of fluorescence anisotropy as mentioned in Procedure Section. Figure shows average anisotropy values of 12 patients with standard error.

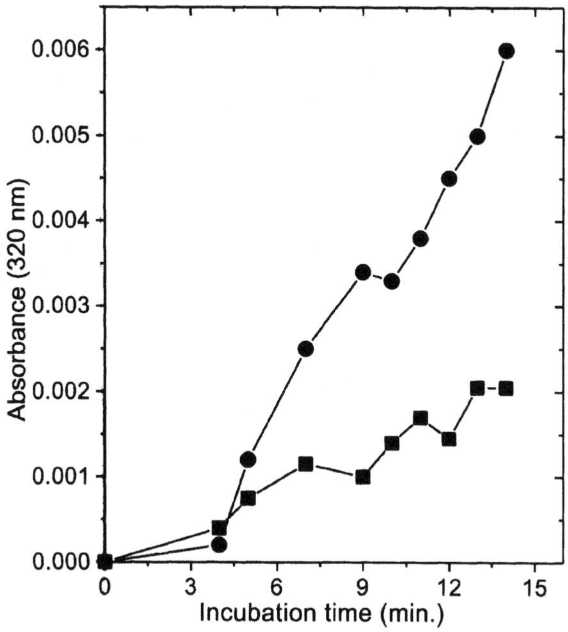

Figure 2. Absorbance of adrenochrome in presence of SOD. Epinephrine was mixed with carbonate buffer (pH 10) containing cell lysate (2.9 μg proteins) followed by incubation at 25 °C. The absorbance of samples was measured at different time of incubation at 320 nm. Control (■) and irradiated (●).

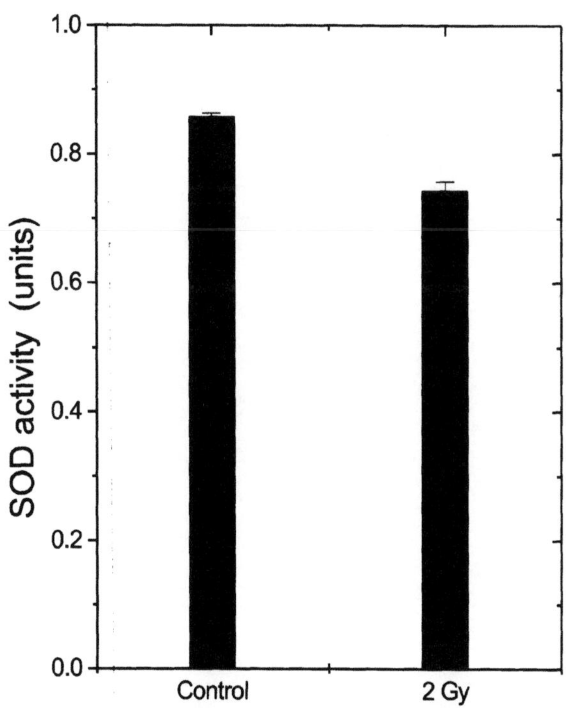

Figure 3. SOD activity in cervical carcinoma after radiation treatment. Lysate of unirradiated and irradiated was mixed with epinephrine in reaction condition. The absorbance was determined after 5 min incubation at 25 °C. The activity of SOD in samples was calculated in terms of units as mentioned in Ref. [10].

Table 1. Percentage apoptosis in cervical cancer cells after radiotherapy. The biopsy samples were obtained from pre- and post-irradiated (2 Gy) cancer patients and cell suspension was prepared in DMEM. The percentage apoptosis in control and irradiated samples was determined by annexin-V and TUNEL method as protocol provided along the kit. The percent apoptosis is average of 9 patients with their respective standard error values

	Apoptosis (%)	
	Annexin-V	TUNEL
Control	38.5 ± 0.2	14.8 ± 4.0
2 Gy	59.4 ± 1.1	28.2 ± 7.7

V protocol was significantly higher than TUNEL method, which requires further studies but probably reflects sensitive detection of membrane surface changes following irradiation. Preliminary fluorescence microscopic examination and flow cytometric studies on γ irradiation induced apoptosis in thymocytes using Annexin V and propidium iodide labeling have shown a sensitive detection of radiation induced membrane associated changes preceding nuclear damage (Our unpublished data). However, these studies are required to be confirmed in cervical carcinoma cells.

Notes on suppliers

1. Sigma Chemical Company, St. Louis, USA
2. GIBCO BRL, Grand Island, N. Y. 14072, USA
3. Roche Diagnostics Corporation, PO Box. 50414, Indianapolis, USA

References

1. Bowen ID, Bowen SM (1990). Programmed cell death in tumours and tissues. Cambridge: University Press.
2. Vermes I et al. (1995). A novel assay for apoptosis flow cytometric detection of phosphatidylserine expression on early apoptotic cells using fluorescein labeled annexin-V. J Immun Meth 184: 39–51.
3. Halliwell B, Gutteridge JMC (1989). Free radical in biology and medicine. Oxford: University Press.
4. Pandey BN, Mishra KP (1999). Radiation biology for cancer radiotherapy. Adv Radiat Biol & Peace UP J Zool Soc Suppl. 2: 45–51.
5. Benderitter M, Vicent-Genod L, Berroud A et al. (1999). Radio-induced structural membrane specifications: A potential bioindicator of ionizing radiation exposure? Intl J Radiat Biol 75: 1043–1053.
6. Pandey BN, Mishra KP (1999). Radiation induced oxidative damage modification by cholesterol in liposomal membrane. Radiat Phys Chem 54: 481–489.
7. Pandey BN, Mishra KP (2000). Fluorescence and ESR studies on membrane oxidative damage by gamma radiation. Appl Mag Reso 18: 483–492.

8. Hartel S, Ojeda F, Diehl H (1998). Cholesterol induced variations of membrane dynamics related to the induction of apoptosis in mouse thymocytes. Intl J Radiat Biol 74: 607–615.
9. Lowry OH, Rosebrough NJ, Lewis Farr A, Handall RJ (1951). Protein measurement with the folin phenol reagent. J Biol Chem 193, 265–275.
10. Sun M, Zigman S (1978). An improved spectrophotometric assay for superoxide dismutase based on epinephrine autoxidation. Anal Biochem 90: 81–89.
11. Ojeda F, Diehl H, Folch H (1991). Radiation induced membrane changes and programmed cell death: Possible inter-relationships. Scann Microsc 8: 615–651.
12. Ramakrishnan N, McClain DE, Catravas GN (1993). Membrane as a sensitive target in thymocyte apoptosis. Int J Radiat Biol 63: 693–701.
13. Mates JM, Perez-Gomez C, Castro IND (2000). Antioxidant enzymes and human diseases. Clin Biochem 32: 595–603.

Address for correspondence: Dr K. P. Mishra, Radiation Biology and Health Sciences Division, Bhabha Atomic Research Centre, Mumbai 400 085, India
Phone: +91-22-559 5043/47/5505217;
Fax: +91-22-550 5151
E-mail: kpm@magnum.barc.ernet.in;
mishra_kaushala@rediffmail.com)

Methods in Cell Science **24**: 69–76 (2002).

Nitric oxide mediated modulation of free radical generation response in the rat polymorphonuclear leukocytes: A flowcytometric study

Madhu Dikshit & Prashant Sharma
Division of Pharmacology, Central Drug Research Institute, Lucknow-226001, India

Abstract. Nitric oxide (NO) synthesis and free radical generation from polymorphonuclear leukocytes (PMNs) play an important role in several pathological conditions. It is therefore important to understand the regulatory mechanisms of free radical generation from PMNs. Flowcytometry can be used to assess generation of reactive oxygen and nitrogen species from PMNs by using fluorescent probes. In the present study regulation of NO synthesis in the control and lipopolysaccharide (LPS) treated rat PMNs has been investigated. Free radical generation was assessed by flow cytometry using a dye, 2′7′-dichlorodihydrofluorescein diacetate (DCFDA), dihydrorhodamine-123 (DHR) and 4,5-diaminofluorescein diacetate (DAF). Superoxide dismutase (SOD), and catalase significantly attenuated the arachidonic acid (AA, 1×10^{-6} M) induced free radical generation, while 4-aminobenzoicacid hydrazide (ABH), myeloperoxidase (MPO) inhibitor had no significant effect. Intracellular and extracellular calcium levels also modulated FR generation. AA induced free radical generation from PMNs was also enhanced significantly after LPS treatment. NO synthase (NOS) inhibitors, aminoguanidine (AG) and 7-nitroindazole (NI) inhibited arachidonic acid induced free radical generation from LPS treated PMNs, while in control PMNs NOS inhibition had no effect. Augmentation of free radical generation from rat PMNs following LPS treatment seems to be regulated by NO.

Key words: Calcium, Nitric oxide, Polymorphonuclear leukocytes, Dihydrorhodamine-123, 2′7′-Dichlorodihydrofluorescein diacetate, Reactive oxygen species

Abbreviations: ABAH = 4-Aminobenzoic acid hydrazide; AG = Aminoguanidine; AA = Arachidonic acid; CPA = Cyclopiazonic acid; DCFDA = 2′,7′Dichlorodihydrofluorescein diacetate; DHR = Dihydrorhodamine-123; DAF = 4,5-Diaminofluorescein diacetate; 4-400 = Dowex 50WX; 1077 and 1119 = Histopaque; LPS = Lipopolysacharide; 7NI = 7-Nitroindazole; L-NAME = L-Nitroarginine methyl ester; TMB-8 = 8-(N.N-Diethylamino) octyl 3,4,5-trimethyl benzoate; PMA = Phorbol myristate acetate; SOD = Superoxide dismutase; OZ = Opsonised zymosan

1. Introduction

PMNs represent almost 50–60% of the total circulating leukocyte population and migration of these cells from blood to tissues invaded by microorganisms is a key feature of the protective host inflammatory response [1, 2]. Apart from PMNs role of NO in inflammatory response is also well recognized [3]. The enzyme responsible for ROS generation is a complex multi-component enzyme – NADPH-oxidase, that has both cytoplasmic and membrane domains. In the dormant PMNs, NADPH-oxidase remains unassembled, but in the activated PMNs both domains of NADPH-oxidase assemble and increase the oxygen consumption to generate reactive oxygen species (ROS).

Lipopolysaccharide (LPS), a component of the outer membrane of gram negative bacteria, upregulates the production of cytokines in various cell types including PMNs [4]. Excessive production of NO in inflammation occurs by the upregulation of inducible nitric oxide synthase (iNOS) in response to inflammatory cytokines and LPS [5]. LPS also primes PMNs for augmented free radical generation [2]. The present study was undertaken to understand the NO mediated regulation of free radical generation in normal and LPS treated PMNs.

2. Materials

4-Aminobenzoic acid hydrazide (ABH), aminoguanidine (AG), arachidonic acid (AA), L-arginine, catalase, cyclopiazonic acid (CPA), 2′,7′dichlorodihydrofluorescein diacetate (DCFDA), dihydrorhodamine-123 (DHR) and 4,5-diaminofluorescein diacetate (DAF), Dowex 50WX (4-400), Histopaque (1077 and 1119), lipopolysaccharide (LPS), 7-nitroindazole (7NI), L-nitroarginine methyl ester (L-NAME), 8-(N.N-diethylamino) octyl 3,4,5-trimethyl benzoate (TMB-8), phorbol myristate acetate (PMA), superoxide dismutase (SOD), zymosan particles were purchased from Sigma Chemicals (USA). Antibodies to CD45-PE, and CD11b-FITC were procured from Becton Dickinson (USA). L-[^{3}H] Arginine was procured from Amersham (UK), while all other chemicals of analytical grade were procured from SRL, Mumbai (India).

3. Procedures

Male Sprague Dawley rats (150–180 g) were obtained from the National Laboratory Animal House of the Institute. They were housed in polypropylene cages and provided with chow pellets and water *ad libitum*. All animal experiments were conducted in accordance to the ethical guidelines of the institute. Rats were pretreated with LPS, 1 mg/kg (i.p) for 4 hours.

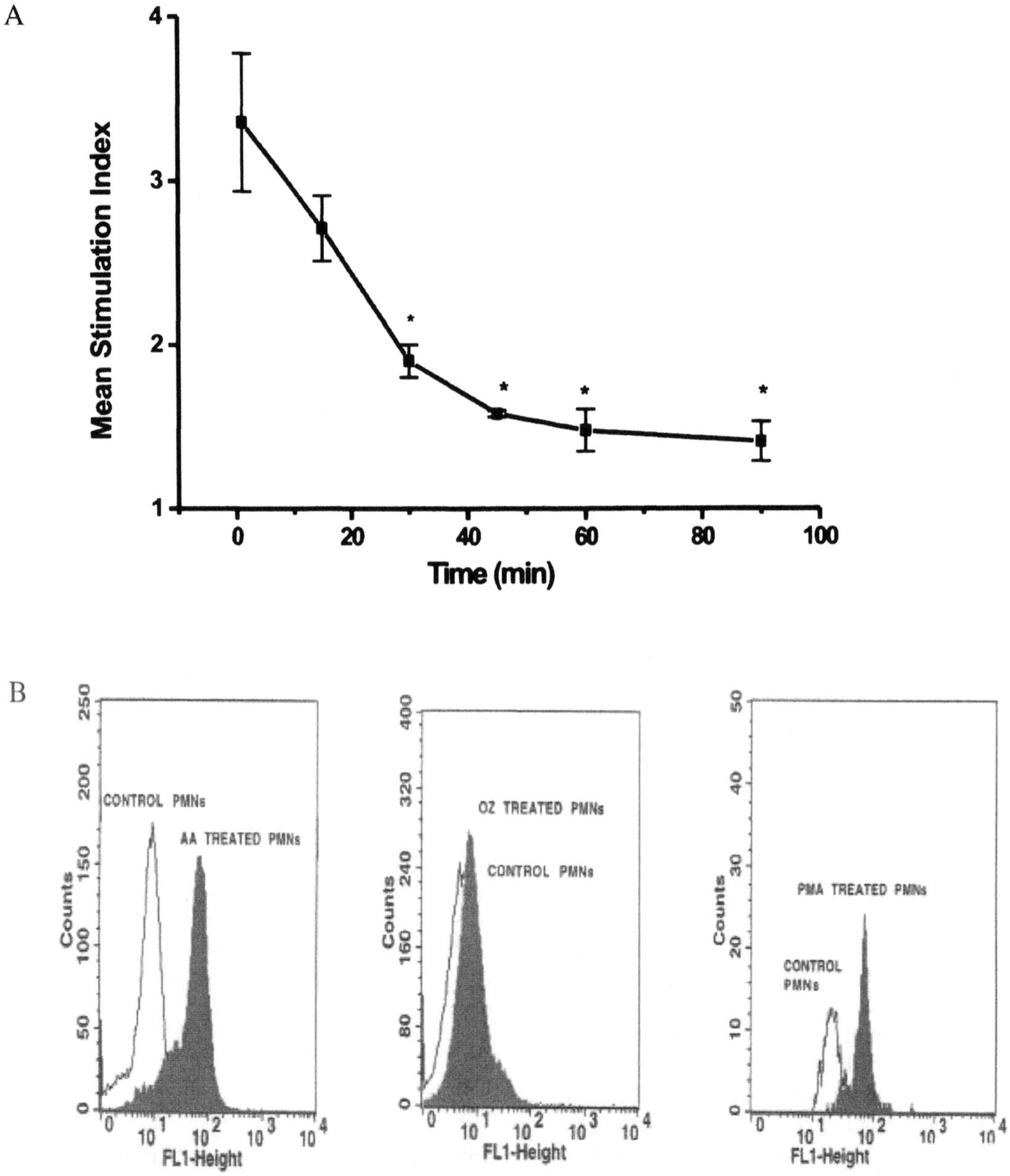

Figure 1. (A) Time dependent arachidonic acid induced free radical generation. 5 × 10^5 cells were loaded with 10 μM of DCFDA for 15 min at 37 °C. ROS generation was induced by AA (3 × 10^{-6} M) and samples were subsequently evaluated at different time periods on FACS Calibur flow cytometer by gating the PMNs population. **p* < 0.001 in comparison to free radical generation after 15 min. (B) Representative histogram tracings showing effect of AA, OZ and PMA on free radical generation. PMNs (5 × 10^5 cells) loaded with either 10 μM of DCFDA or 5 μM DHR and then treated with AA (1 × 10^{-6} M) for 15 min., OZ (1.25 μg) for 30 min or PMA (3 × 10^{-9} M) for 30 min. Subsequently the samples were cooled on ice and evaluated on a FACS Calibur flow cytometer by gatting the PMNs population.

3.1. *Isolation of PMNs*

Rat blood was collected by cardiac puncture in sodium citrate (0.129 M, pH 6.5, 9:1 v/v) under ether anesthesia. PMNs from blood were isolated from the control animals as well as from the rats 4 hours after LPS treatment by the method of Boyum [6]. Platelet rich plasma was removed by centrifugation at 250 g for 20 min at 20 °C (Sigma Centrifuge, Germany) and the buffy coat was subjected to dextran sedimentation and Histopaque gradient as described earlier [7].

Cells after counting were suspended (1×10^6 cells/ml) in the Hank's Balanced Salt Solution (HBSS, Composition (mM): NaCl 138; KCl 2.7; Na_2HPO_4 8.1; KH_2PO_4 1.5; Glucose 10; pH 7.4 and their viability was tested by Trypan blue exclusion test, which was never less than 95%.

3.2. *Flow cytometric analysis of free radical generation*

Purity of PMNs preparation was evaluated by labeling the PMNs with anti-rat CD45-PE and anti-rat CD11b-FITC (Becton Dickinson, USA). Cells were labeled according to the manufacturer's instruction. PMNs (5×10^5 cells) were equilibrated in 850 µl of HBSS with 50 µl of DHR (dihydrorhodamine-123, 5 µM), DCFDA (10 µM) or DAF (10 µM) for 10–15 min at 37 °C. ROS generation was induced by arachidonic acid (AA, 1×10^{-6} M) and the cells were incubated for 15 min at 37 °C. In some experiments cells were incubated with AA for various time (30, 45, 60, 90 min) intervals. After the incubation time samples were cooled on ice and evaluated on a FACS Calibur flow cytometer (Becton-Dickinson, USA). FL1 fluorescence of 10,000 cells was collected in each experiment by gating and acquiring the PMNs population [8, 9]. Free radical generation using DHR and DCFDA was evaluated by Cell Quest program.

To study the effect of various interventions on free radical generation, PMNs were preincubated with ABAH (100 µM) for 15 min, CPA (100 µM) for 5 min, TMB-8 (50 µM) for 5 min, SOD (100 U) and catalase (500 U) for 5 min., 7-NI (1 mM) or AG (1 mM) for 1 hr at 37 °C. Cells were loaded either with 5 µM DHR or 10 µM DCFDA at the end of the incubation period and were further incubated for 15 min at 37 °C. ROS generation from PMNs was induced by AA (1×10^{-6} M). OZ (1.25 µg) or PMA (3×10^{-9} M) was also used for the free radical production from rat PMNs. Free radical generation from PMNs has been reported as stimulation index, which is the ratio of the mean fluorescence of the stimulated and unstimulated cells as described earlier [10].

3.3. *NOS estimation*

PMNs (1×10^7 cells) suspended in incubation buffer (Composition (mM): HEPES 25; NaCl 140; KCl 5.4; $CaCl_2$ 1; $MgCl_2$ 1; pH 7.4) were incubated with L-[^{3}H]-arginine (0.1588 µM) for 30 min at 37 °C and the reaction was stopped by the addition of ice cold stop buffer (Composition (mM): NaCl 118; KCl 4.7; KH_2PO_4 1.18; $NaHCO_3$ 1; EDTA 4; L-NAME 2; pH

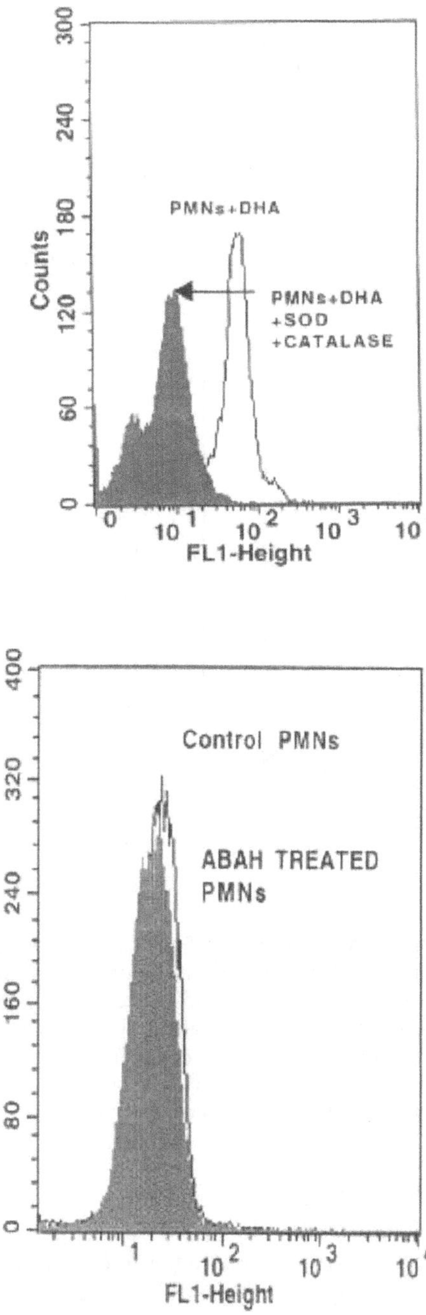

Figure 2. (A) AA (1×10^{-6} M) induced free radical generation in the presence of superoxide (SOD, 100 Units/ml) and Catalase (500 Units/ml). (B) Effect of MPO inhibitor, ABH (100 µM) on AA induced free radical generation from PMNs.

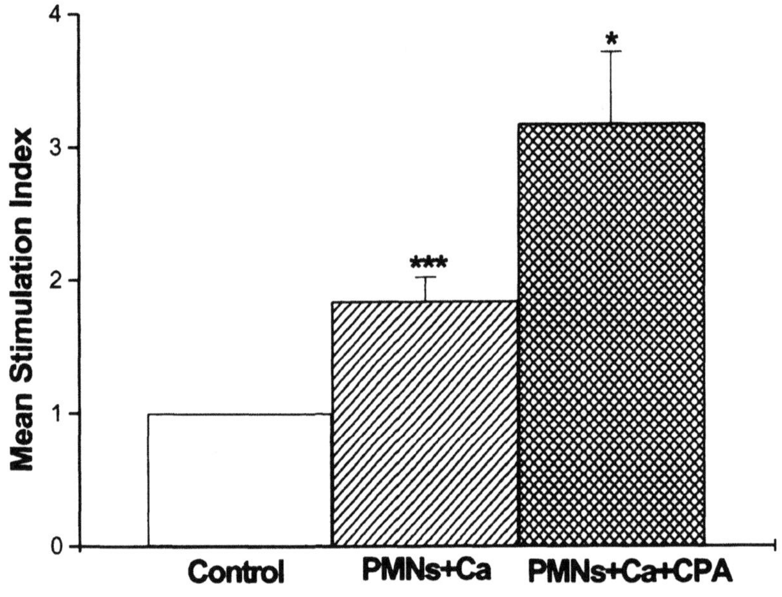

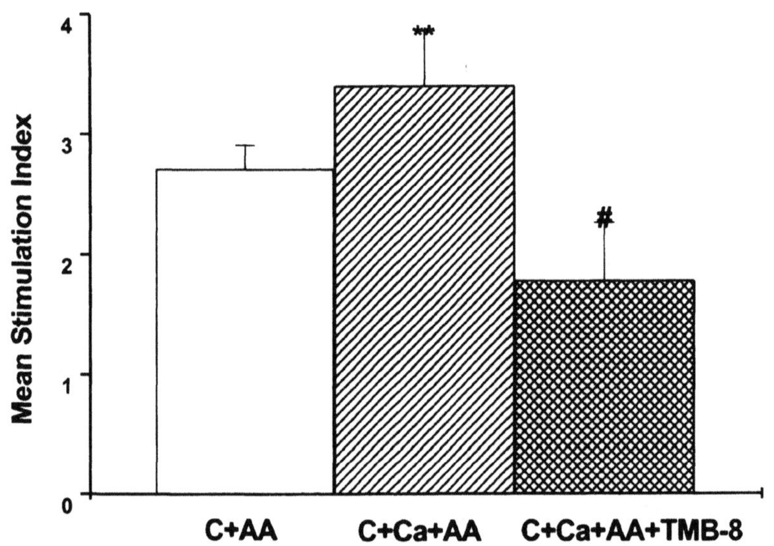

Figure 3. (A and B) Effect of calcium and calcium modulators on free radical generation. PMNs, suspended in calcium containing buffer were pretreated with calcium modulators (CPA [100 μM], or TMB-8 [50 μM]) for 5 min and loaded with 10 μM of DCFDA for 15 min at 37 °C. Free radical generation was measured in both control and AA stimulated PMNs for 15 min by FACS Calibur flowcytometer by gating the PMNs population. $***p < 0.001$ in comparison to unstimulated control, $*p < 0.05$ in comparison to basal free radical generation in PMNs suspended in calcium containing buffer, $**p < 0.05$ in comparison to free radical generation in PMNs suspended in calcium free medium, $^{\#}p < 0.05$ in comparison to free radical generation in stimulated PMNs suspended in calcium containing buffer.

5.5). The cells were sonicated and the mixture was passed through the Dowex (Na^+ exchanger) columns. The radioactivity present in the eluent was estimated with the help of a β-scintillation counter (LKB-Pharmacia). In addition, to evaluate the effect of NOS inhibitors on NOS activity, PMNs were incubated with aminoguanidine (AG, 1 mM) or 7-nitroindazole (7-NI, 1 mM) for 1 hr at 37 °C prior to the addition of radioactive arginine. NOS activity in the PMNs has been reported as (pmoles/30 min/10^7 cells).

3.4. *Statistical analysis*

Results have been expressed as mean ± SE of at least 5 experiments in each group. Comparison between the different groups was performed by Students '*t*' test and differences were considered significant at $p < 0.05$.

4. Results and discussion

4.1. *Free radical generation from control PMNs*

Migration of PMNs from blood to the specific tissue following a pathological insult is a key feature of the host inflammatory response, which is often localized and protective to eliminate the invading microbes. However, damaging role of PMNs derived free radicals has been suggested in the reperfusion injuries. Free radicals are formed in PMNs due to the activation of NADPH-oxidase. In the present study generation of free radicals from normal and LPS treated PMNs has been assessed by flowcytometry.

Mean stimulation index of AA (1×10^{-6} M) induced free radical generation in the control rat PMNs was 2.71 ± 0.2 ($n = 37$), which decreased in a time dependent manner (Figure 1A). Reduction was not significant at 15 min while, at 30, 60 and 90 min it was significantly reduced. Most of the experiments in the present study were therefore performed during the first 15 min of incubation with AA. Control rat PMNs also exhibited opsonised zymosan (OZ), or PMA induced free radical generation (Figure 1B). SOD and catalase significantly inhibited the AA induced free radical generation response (Figure 2A). MPO inhibitor, ABAH had no significant effect on AA induced free radical generation as estimated by DCFDA or DHR (Figure 2B).

The dyes, DHR and DCFDA were used in the present study to measure free radical production from PMNs by flow cytometry [9, 10, 11, 12]. Both diffuse into the cells and are rapidly oxidized to highly fluorescent rhodamine 123 or DCF primarily by hydrogen peroxide, and the reaction is markedly enhanced by peroxidase [13, 14]. DCFDA in addition to H_2O_2, is also oxidized by NO, nitrogen dioxide

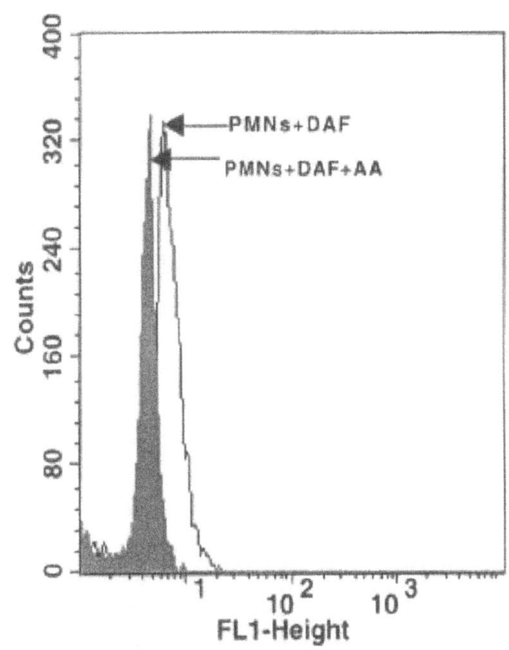

Figure 4. Nitric oxide generation in the resting and AA (1×10^{-6} M) stimulated PMNs. Cells were loaded with DAF (10 μM) for 10–15 min and then were treated with AA.

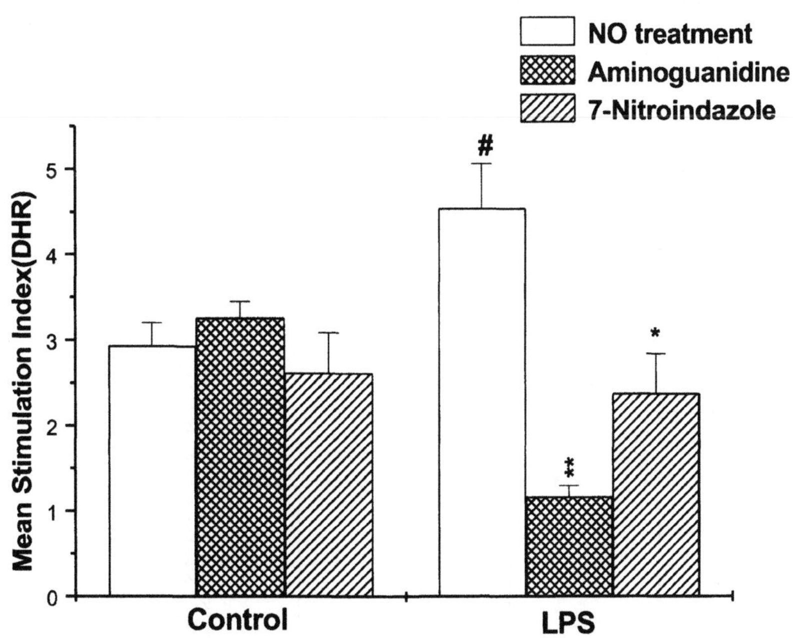

Figure 5. Effect of aminoguanidine or 7-nitroindazole on the AA induced free radical generation from control and LPS treated PMNs. PMNs were preincubated with 7-NI (1 mM) or with AG (1 mM) for 1hr at 37 °C. After the incubation period cells were loaded with 5 μM DHR ($n = 6$) and were incubated for 10 min at 37 °C. ROS generation was induced by AA (1×10^{-6} M) and has been reported in terms of stimulation index. *$p < 0.05$; **$p < 0.01$ in comparison to their respective untreated controls. $^{\#}p < 0.02$, in comparison to the control.

($\cdot NO_2$), HOCl and peroxynitrite a reaction product of O_2^- and NO [15, 16] while DHR-123 in addition to H_2O_2 is also oxidized by peroxynitrite [17]. Since the dyes used react with reactive nitrogen species apart from ROS, NOS inhibitors were used to further investigate the involvement of NO in free radical generation.

Control cells exhibited AA induced free radical generation by using DCFDA and DHR-123. Free radical scavenging enzymes, SOD and catalase significantly inhibited the free radical generation suggesting that superoxide radical and hydrogen peroxide are formed following AA addition. While AA induced free radical generation was not attenuated in presence of ABAH, a MPO inhibitor, indicating that MPO is not a major participant in free radical formation being measured by DCFDA or DHR-123.

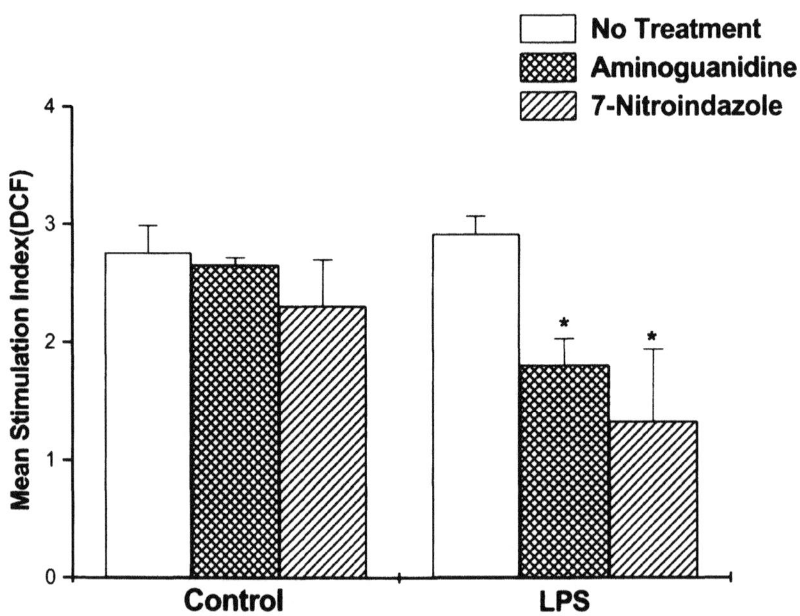

Figure 6. Effect of 7-nitroindazole or aminoguanidine on the AA induced free radical generation from control and LPS treated PMNs. PMNs were preincubated with 7-NI, 1mM or with AG, 1mM for 1hr at 37 °C. After the incubation period cells were loaded with 10 μM DCFDA ($n = 3$) and were incubated for 10 min at 37 °C. ROS generation was induced by AA (1×10^{-6} M) and has been reported in terms of stimulation index. *$p < 0.02$ in comparison to their respective untreated controls.

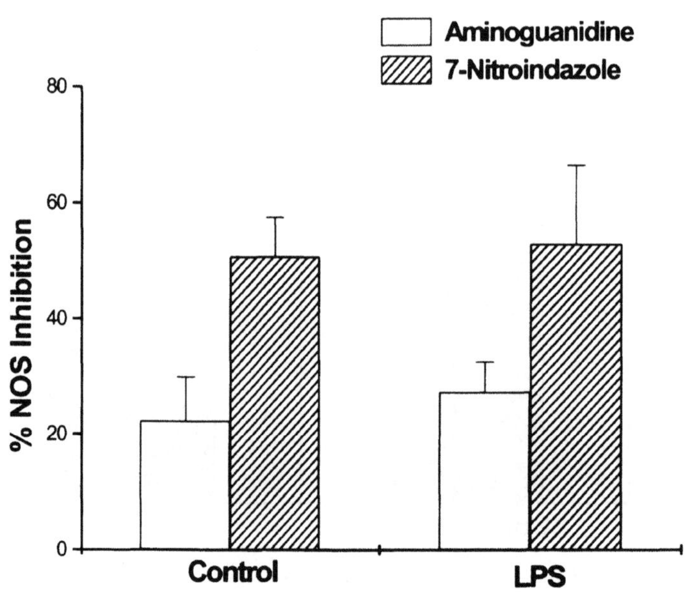

Figure 7. Effect of NOS inhibitors on the PMNs NOS activity. To evaluate the effect of NOS inhibitors on NOS activity, PMNs were incubated with aminoguanidine (1 mM) or 7-nitroindazole (1 mM) for 1 hr at 37 °C prior to the addition of radioactive L-arginine. NOS activity in the PMNs has been reported as (pmoles/30 min/10^7 cells).

To investigate regulation of free radicals generation in PMNs by calcium, effect of calcium modulators was simultaneously investigated in both control and AA stimulated PMNs. PMNs suspended in calcium (1 mM) containing buffer exhibited significant enhancement in free radical generation in comparison to PMNs suspended in calcium free buffer and addition of AA further augmented the free radical generation response (Figure 3). To understand further the contribution of calcium in free radical generation, PMNs, suspended in calcium containing buffer were treated with CPA or TMB-8. CPA, which inhibits microsomal Ca^{+2}-ATPase and protrude the store-regulated calcium influx (18) significantly enhanced the free radical generation in PMNs (Figure 3), while TMB-8 acts on membrane to inhibit the influx of extracellular calcium and on intracellular storage sites to block the release of calcium (19) significantly reduced the free radical generation from PMNs even though PMNs were stimulated with AA (Figure 3) These results suggest that calcium influx from extracellular space or release of calcium from intracellular pools play an important role in the generation of free radicals from PMNs. The respiratory burst response in PMNs is thus regulated by extracellular and intracellular calcium [20].

Circulating PMNs from rat and human [21] possess mRNA and the nNOS protein. NO has been found to be involved in modulating ROS generation from PMNs [7]. PMNs exhibited spontaneous release of NO in resting condition, while addition of AA (1×10^{-6} M) to cell suspension results in significant decrease in the DAF florescence (Figure 4). NOS inhibitors had no significant effect on AA induced free radical generation as assessed by DCFDA or DHR (Figures 5, 6 and 7). Thus it appears that in the control cells NO is not participating in the free radical generation response.

4.2. *Effect of LPS treatment*

LPS elicits a variety of responses in PMNs such as receptor upregulation [22], actin assembly [23], and priming of the cells to release superoxide radicals [2, 24]. It has been found that administration of LPS in rats causes priming of NADPH-oxidase activity by inducing the translocation of flavocytochrome b555 from specific granules to the plasma membrane together with limited translocation of p47 phox [2].

Mean stimulation index, a ratio of free radical generation from AA stimulated PMNs was significantly more in the LPS treated rat PMNs in comparison to the control PMNs (Figures 5 and 6). In LPS treated cells response to OZ and PMA were also augmented (Figure 8). Both AG and 7-NI, significantly inhibited AA induced free radical generation in the LPS treated cells (Figures 5 and 6).

NOS activity in the PMNs was significantly inhibited in presence of 7-NI and AG in the control and LPS treated cells (Figure 7). Free radical generation was however, reduced significantly only in the LPS treated cells in presence of NOS inhibitors, while in the control cells NOS inhibition had no significant effect on the stimulation index, a measure of free radical generation. Thus it appears that in control cells concentration of NO is not sufficient enough to generate reactive nitrogen species. While induction of NO pathway in the PMNs following LPS treatment augments NO generation, which seems to participate in the free radical generation response.

Results of the present study suggest an interrela-

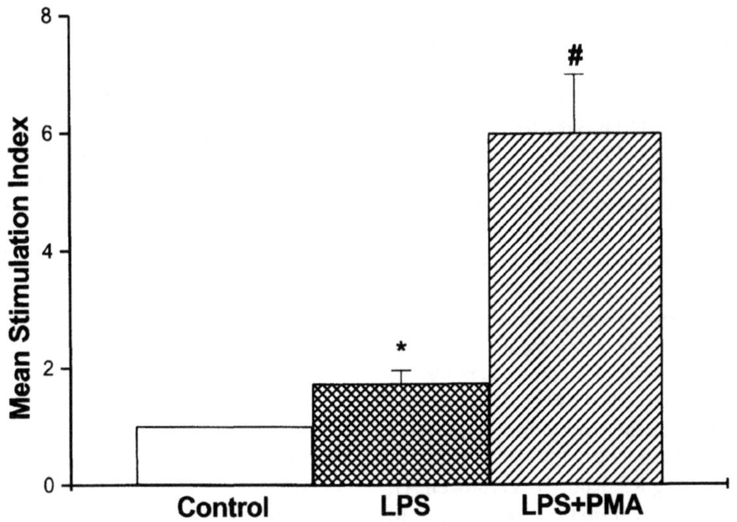

Figure 8. Effect of PMA on basal free radical generation from LPS treated rat PMNs. PMNs isolated from LPS treated rats were loaded with 5 μM DHR and incubated with PMA (3×10^{-9} M) for 30 min. Samples were subsequently cooled on ice and evaluated on a FACS Calibur flow cytometer by gating the PMNs population. *$p < 0.01$ in comparison to free radical generation in control PMNs. #$p < 0.01$ in comparison to free radical generation in LPS treated PMNs.

tionship between NO synthesis and free radical generation in LPS treated PMNs.

Acknowledgement

The author wish to acknowledge the excellent technical assistance of Mr. A.L.Vishwakarma in flow cytometric studies.

References

1. Clark RA (1990). The human neutrophil respiratory burst oxidase. J Infect Dis 161: 1140–1147.
2. Condliffe AM, Kitchen E, Chilvers ER (1998). Neutrophil priming: pathophysiological consequences and underlying mechanisms. Clin Sci 94: 461–471.
3. Miles AM, Owens MW, Mulligan S, Johnson GG, Fields JZ, Ing TS, Kottapali V, Keshavarazian A, Grisham MB (1995). Nitric oxide synthase in circulating vs. extravasated polymorphonuclear leukocytes. J Leukocyte Biol 58: 616–622.
4. Haziot A, Tsuberi BZ, Goyert SM (1993). Neutrophil CD14: Biochemical properties and role in the secretion of tumour necrosis factor α in response to lipopolysaccharide. J Leukocyte Biol 150: 5556–5565.
5. Kolls J, Yie J, Le Blanc R, Melinski T, Nelson S, Summer W, Greenburg SS (1994). Rapid induction of messenger RNA for nitric oxide synthase in rat neutrophils *in vivo* by endotoxin and its suppression by prednisolone. Proc Soc Exp Biol Med 205: 220–229.
6. Boyum A (1968). Separation of leukocytes from blood and bone marrow. Scand. J Clin Lab Invest 21 (suppl 97): 77–89.
7. Sethi S, Singh MP, Dikshit M (1999). Nitric Oxide mediated augmentation of polymorphonuclear free radical generation after hypoxia-reoxygenation. Blood 93: 333–340.
8. Szucs S, Vamosi G, Paka R, Sarvary A, Bardoo H, Balazs M, Kappelmayer J, Toth L, Szollose J, Single AR (1998). Single cell measurement of superoxide anion and hydrogen peroxide production by human neutrophils with digital imaging fluorescence microscopy. Cytometry 33: 19–31.
9. Bass DA, Olbrantz P, Szejda P, Seeds MC, McCall CE (1986). Subpopulation of neutrophils with increased oxidative product formation in blood of patient with infection. J Immunol 136: 860–866.
10. Vowells SJ, Sekhsaria S, Malech HL, Shelit M, Fleisher TA (1995). Flow cytometric analysis of the granulocyte respiratory burst; a comparative study of fluorescence probes. J Immunol Methods 178: 89–97.
11. Henderson LM, Chappell JB (1993). Dihydrorhodamine 123: A fluorescent probe for superoxide generation. Eur J Biochem 217: 973–980.
12. Pelt LJ, Zwiter R, Wenlig RS, Roos D, Verhowen AJ, Bolscher BGJM (1996). Limitation of the use of dihydrorhodamine 123 for flow cytometric analysis of the neutrophil respiratory burst. J Immunol Methods 191: 187–196.
13. Emmendroff A, Hecht M, Matlher MLL, Roesler J (1990). A fast and easy method to determine the production of reactive oxygen intermediate by human and murine phagocytes using dihydrorhodamine 123. J Immunol Methods 131: 269–275.
14. Bass DA, Parce JW, Dechatelet LR, Szejda P, Seeds MC, Thomas M (1983). Flow cytometric studies of oxidative product formation by neutrophils: a graded response to membrane stimulation. J Immunol 130: 1910–1917.
15. Kooy NW, Royall JA, Ischiropoulos H, Beckman, JS (1994). Peroxynitrite-mediated oxidation of dihydrorhodamine 123. Free Radic Biol Med 16: 149–156.
16. Rao KM, Padmanabhan J, Kilby DL, Cohen HJ, Currie MS, Weinberg JB (1992). Flow cytometric analysis of nitric oxide production in human neutrophils using dichloroflouroscein diacetate in the presence of calmodulin inhibitor. J Leukocyte Biol 51: 496–500.
17. Crow JP (1997). Dichlorodihydrofluoresceine and Dihydrorhodamine 123 are sensitive indicators of peroxynitrite *in vitro*: implication of intracellular measurement of reactive nitrogen species. Nitric Oxide: Biol Chem 1: 145–157.
18. Geiszt M, Kaldi K, Szebereryi JB, Light E (1995). Thapsigargin inhibits calcium entry into human neutrophils granulocytes. Biochem J 305: 525–528.
19. Chiou CY, Malagodi HM (1975). Studies on the mechanish of action of a new Ca^{+2} antagoinst, 8-(N.N-diethylamino) octyl 3,4,5-trimethyl benzoate hydrochloride in smooth and skelton muscles. Brit J Pharmacol 53: 279–285.
20. Bei L, Hu T, Qian ZM, Shen X (1998). Extracellular calcium regulates the respiratory burst of human neutrophils. Biochem Biophys Acta 1404:, 475–483.
21. Greenberg SS, Ouyang J, Zhao XF, Gile JD (1998). Human and rat neutrophils constitutively express neural nitric oxide synthase mRNA. Nitric Oxide: Biol Chem 2: 203–212.
22. Vosebeck K, Tobias P, Mueller H, Allen RA, Arfors KE, Ulevitch RJ, Sklar LA (1990). Priming of polymorphonuclear granulocytes by lipopolysaccharide and its complexes with lipopolysaccharide binding protein and high-density lipoprotein. J Leukocyte Biol 47: 97–104.
23. Avidi NJ, Gerwins P, Johnson GL, Worthen GS (1996). Activation of a p38 mitogen-activated protein kinase in human neutrophils by lipopolysaccharide. J Immunol 156: 4867–4875.
24. Sapira L, Champagne C, Gordon B, Amar S, VanDyke TE (1995). Lipopolysaccharide priming of superoxide release by human neutrophils: Role of membrane CD14 an serum LPS Binding proteins. Inflammation 19: 289–295.

Address for correspondence: Dr. Madhu Dikshit, PhD, Pharmacology Division, Central Drug Research Institute, Lucknow -226001, India
Phone: 91-522-212411 to 18 PABX (4254);
Fax: 91-0522-223405
E-mail: madhudikshit@yahoo.com

Methods in Cell Science 24: 77–78 (2002).

Flowcytometric detection of PNH defect and response to therapy in aplastic anemia patients

Neelam Varma, Subhash Varma, Harpreet Vohra, Kiran Malik & Gurjeevan Garewal
Departments of Hematology, Internal Medicine and Experimental Medicine & Biotechnology, Postgraduate Institute of Medical Education and Research, Chandigarh

1. Introduction

Clonal hemopoietic stem cell disorders, characterized by features of bone marrow failure, include aplastic anemia (AA) and paroxysmal nocturnal hemoglobinuria (PNH) [1]. These diseases differ in the etiopathogenesis and clinical presentations, although treatment options for both still remain a clinician's nightmare. Damashek, way back in 1960s, first appreciated the common link between AA and PNH i.e., one disorder can evolve into the other. The 'PNH defect', whether de novo or secondary (arising in the background of other hematological disorders like AA) is characterized by deficiency or complete lack of more than 20 different membrane proteins, which are covalently attached to the cell membrane via a glycosyl phosphatidyl inositol (GPI) anchor [2]. The absence of GPI anchored complement regulating proteins such as membrane inhibitor of reactive lysis (MIRL or CD 59) and decay accelerating factor (DAF or CD55) is responsible for the phenotype of the disease and same feature is used for diagnostic purposes as well [2]. There is scarcity of data in the literature regarding the frequency and prognostic implications of PNH defect in AA patients [2–4]. Here we elaborate upon our experience on this subject, published earlier as a preliminary communication [4].

2. Material and methods

Ninety adult AA patients were included in this study between March 1996 and November 2001. Tests for PNH (conventional and flowcytometric) were performed at the time of diagnosis and later during the follow up period at 3 to 6 month intervals (2 to 6 times). The diagnosis of AA was made according to the standard criteria, based upon complete blood counts, bone marrow aspiration, and trepline biopsy findings. 'Clinical' and 'laboratory' PNH was diagnosed according to the criteria proposed by Tichelli et al. [5], with the addition of flowcytometry (FCM) parameter for laboratory PNH. Clinical PNH is diagnosed in the presence of hemolytic anemia, overt hemoglobinuria, abdominal or lower back pain, refractory iron deficiency anemia or thrombotic complications. Laboratory PNH is diagnosed in the presence of hemosiderin in the urine and/or Ham's acidified serum test (HAST) and/or sucrose lysis test (SLT) and/or a PNH clone detected by FCM, without clinical features of hemolysis. FCM was performed on the RBCs and granulocytes (of controls and patients) using anti CD55 and CD59 monoclonal antibodies (moAb).

3. Results

The details of the results are depicted in Table 1. None of the AA patients revealed evidence of clinical PNH. At diagnosis, none of the patients showed positive conventional tests, whereas 5 (5.5%) patients were positive on follow up. Using FCM, PNH clone was detected in 8 (8.9%) patients at diagnosis and 28 (31.1%)patients during follow up; more often in granulocytes than RBCs. During the period of study, no adverse effect on clinical outcome was observed vis-à-vis detection of PNH clone in AA. When last seen, 62 patients were in partial remission and 28 were not in remission.

4. Discussion

Aplastic anemia is characterized by decreased blood counts (pancytopenia) and hypocellular bone marrow [1]. PNH patients frequently present with

Table 1. PNH defect in aplastic anemia patients

	At diagnosis	On follow up
Urine hemosiderin	0	2
Hemoglobinuria	0	0
Hemoglobinemia	0	0
HAST*	0	3
SLT*	0	0
CD55/59 deficient		
WBCs	8	28
RBCs	5	13

*Expanded form as in the text.
Conventional tests vs FCM ($p < 0.01$)
Positivity at diagnosis vs at follow up ($p < 0.01$).

intravascular hemolysis, refractory iron deficiency-like blood picture, hypercellular bone marrow and positivity for serological/conventional tests like HAST and SLT (indicating increased sensitivity of RBCs to autologous complement activation) [1, 2]. Since HAST and SLT depend upon deficiency of GPI-anchored proteins on RBCs, the sensitivity of the tests is frequently compromised by the blood transfusions received by AA patients. FCM being a more sensitive technique, circumvents this drawback, quantitates the PNH clone and in addition can detect the PNH defect in WBCs as well. The reported frequency of secondary PNH in AA has varied over the years, primarily due to variable sensitivity of the tests employed [2]. PNH defect is detected in nearly 10–15% AA patients by conventional tests whereas FCM can detect it in 29–52% patients during follow up [2]. We were able to detect a secondary PNH clone in 31.1% of our AA patients, which however did not influence the treatment outcome. Our experience in this regard is similar to only other such study reported earlier [3]. Our results give credence to the hypothesis that PNH clone arises as a hemopoietic escape response to the damaged bone marrow in AA patients. This phenomenon helps in the understanding of biological evolution and inter-relationship of aplastic anemia and PNH disorders.

References

1. Lichtman MA (1995). Classification and clinical manifestations of the hemopoietic stem cell disorders. In: Beutler et al. (eds), Hematology. New York: McGraw-Hill Book Company, pp 229–237.
2. Nishimura J, Murakami Y, Kinoshita T (1999). Paroxysmal nocturnal hemoglobinuria: an acquired genetic disease. American J Hematol 62: 175–182.
3. DeLord C, Tooze JA, Saso R, Marsh JCW, Gordon-Smith EC (1998). Deficiency of glycosyl phosphatidyl inositol-anchored proteins in patients with aplastic anemia does not affect response to immunosuppressive therapy. British J Haematol 101: 90–93.
4. Varma N, Garewal G, Varma S, Vohra H (2000). Flowcytometric detection of PNH defect in Indian patients with aplastic anemia and myelodysplastic syndromes. American J Hematol 65: 264–265.
5. Tichelli A, Grahwohl A, Wursh A, Nissen C, Speck B (1998). Late haematological complications in severe aplastic anemia. British J Haematol 69: 413–418.

Methods in Cell Science **24**: 79–90 (2002).

Host-pathogen interactions: leukocyte phagocytosis and associated sequelae

Jovanka M. Voyich & Frank R. DeLeo
Laboratory of Human Bacterial Pathogenesis, Rocky Mountain Laboratories, National Institute of Allergy and Infectious Diseases, National Institutes of Health, Hamilton, MT 59840, USA

Abstract. Polymorphonuclear leukocytes (PMNs) are a critical component of the human innate immune response and are the first line of defense against invading microorganisms. Phagocytosis of invading microbes induces production of reactive oxygen species (ROS) by PMNs, which facilitates bactericidal activity. In addition to eliminating microorganisms, phagocytosis also accelerates PMN apoptosis, a process critical for resolution of inflammation. Inasmuch as leukocyte phagocytosis and ROS production are key components of the innate immune response, we developed flow cytometric methods to evaluate these processes in human PMNs. In contrast to traditional microscopy-based analyses, the methods described herein provide objective and high throughput measures of host cell-pathogen interactions. Importantly, they can be adapted for use with a number of fluorometric probes, and bacterium and host cell of choice, and each is based upon a common phagocytosis assay system. We also describe methods to measure phagocytosis-induced PMN apoptosis with this assay system. These methods entail detecting surface-exposed phosphatidylserine (early apoptosis), and measuring PMN chromatin condensation and DNA fragmentation (late apoptosis). Taken together, these assays provide rapid and accurate assessment of critical PMN processes.

Key words: Apoptosis, Flow cytometry, Neutrophil, Phagocytosis, Respiratory burst

Abbreviations: DHE = dyhydroethidine; GAS = Group A *Streptococcus*; ROS = reactive oxygen species; PI = propidium iodide; PS = phosphatidylserine; TUNEL = terminal deoxynucleotidyltransferase dUTP nick end labeling; DHR = dihydrorhodamine 123; DCF = 2′7′dicholrofluorescin

1. Introduction

Phagocytic leukocytes are the first line of defense against invading microbes. Polymorphonuclear leukocytes (PMNs) comprise 60–70 percent of all human white blood cells and provide rapid defense against invading pathogens. Therefore, elucidation of specific features of PMN-pathogen interactions is an important step toward understanding infectious diseases and the inflammatory response in humans. The assays described in this Chapter are designed to measure specific phenomena associated with PMN-pathogen interactions and can be adapted for use with other cells of the immune system.

PMNs in circulation survey the body for invading microorganisms, in part, by interacting directly with the vascular endothelium. During infection, host cells and microorganisms in infected tissues produce chemoattractants, such as IL-8 and lipopolysaccharide, respectively [22]. These pro-inflammatory molecules facilitate recruitment of PMNs to infected tissues in an orchestrated molecular cascade of events, which includes up regulation of host cell adhesion molecules, subsequent binding of PMNs to the endothelium, and PMN extravasation through the vessel wall to the site of infection [22]. Binding and phagocytosis of pathogens by human PMNs is promoted either directly or indirectly. PMNs utilize relatively invariant pathogen surface structures such as lipopolysaccharide or peptidoglycan to facilitate direct phagocytosis [22, 34]. Indirect phagocytosis is facilitated by opsonization (coating) of pathogens predominantly with serum complement or specific antibody, although other well-known opsonic host proteins contribute to the process as well [34]. Following phagocytosis, PMNs destroy microorganisms with antimicrobial proteins and reactive oxygen species (ROS), such as superoxide, hydrogen peroxide, and hypochlorous acid [22]. Although PMNs can eliminate most bacterial pathogens, many species have developed clever strategies to evade the innate immune response of the human host [8]. For example, *Yersinia* sp. inhibit the Rho family of GTPases, thereby blocking actin polymerization, a necessary step for phagocytosis [34]. Group A *Streptococcus* (GAS) is a Gram-positive bacterium especially successful at evading PMN phagocytosis and/or the resulting outcomes of PMN phagocytosis [6].

Programmed cell death or apoptosis is an important regulatory mechanism that facilitates normal PMN turnover and resolution of inflammation [24, 31, 32, 39]. PMNs undergo apoptosis spontaneously as part of normal cell turnover in humans [31, 32]. However, PMN apoptosis is accelerated significantly following phagocytosis [14], presumably as a mechanism to facilitate resolution of disease and associated inflammation. For example, apoptosis is induced following phagocytosis of *Mycobacterium tuberculosis* to mediate clearance of bacilli-containing cells, ultimately aiding the host response [25]. Alternatively, some organisms exploit apoptosis to eliminate cells of the immune system [9, 10], or to create an environment of reduced inflammation that is tolerable to the bacterium [38]. Whether induction of PMN apoptosis by phagocytosis is beneficial to the host or pathogen is likely dependent on the species of invading microorganism.

Although significant progress has been made, aided by a new era of genomics, mechanisms employed by bacterial pathogens to alter PMN phagocytosis and associated sequelae are incompletely characterized. Therefore, measuring host-pathogen interactions with high-throughput assays is critical for implementing the extensive data obtained through genomics strategies. Host-pathogen interactions are complex and, therefore, demand multi-parameter analyses. This Chapter describes procedures for relative quantitation of adherence and phagocytosis of live bacteria by human PMNs. Additional protocols outline multi-parameter analysis of oxidative burst in association with phagocytosis. The Chapter concludes with assays that determine phagocytosis-induced PMN apoptosis. In summary, relative quantitation of specific features of host-pathogen interactions can be determined rapidly and concomitantly from a large number of cells in a single sample using flow cytometry.

Protocol 1: Measuring adherence and phagocytosis of bacteria with human PMNs

Determining phagocytosis is critical to understanding PMN function and bacterial pathogenesis. Electron microscopy [28], fluorescence combined with phase-contrast microscopy [2], and confocal laser-scanning microscopy have been employed successfully [33, 7]; however, these microscope-based techniques are tedious, time consuming, and the ability to quantitate extensive numbers of cells is limited. Microscope-based phagocytosis assays also involve subjective data interpretation. Flow cytometry provides an objective, high throughput alternative for the measure of phagocytosis and can be used to track kinetics of ingestion [16, 17, 21]. This methodology can be applied to screen opsonic activity of human sera specific to a given bacterial species, and provide high-throughput screening of vaccine efficacies [1, 12, 20, 26, 27, 37]. Importantly, flow cytometry provides the ability to measure changes in cell surface receptors and activation markers concurrently with phagocytosis [17, 35].

The assay described below synchronizes PMN phagocytosis by centrifugation and temperature [7]. Synchronization of the assay allows for better assessment of phagocytosis kinetics, since nearly all of the bacteria are either in direct contact with or in close proximity to PMNs. The assay is based on the ability of Trypan Blue to quench signal from fluorescently labeled bacteria and allows rapid quantitation of PMNs with associated bacteria and those containing internalized bacteria. The entire procedure takes approximately 4–6 hours, including PMN isolation and specific manipulations with bacteria, e.g., opsonization or heat killing.

2a. Materials

Dulbecco's Phosphate Buffered Saline (DPBS) (Sigma D-8662).[1]
Purified human PMNs (Support Protocol 1).
RPMI 1640 medium (Invitrogen/Life Technologies)[2] containing 10 mM HEPES (Sigma H-0887)[1] (RPMI-H).
96 well flat-bottom plates (Fisher Scientific).[3]
Fluorescein isothiocyanate solution (FITC, Sigma F-4274).[1]
Trypan blue solution (J.T. Baker X242-03, Fisher Scientific)[3], 2 mg/ml in 0.02 M citrate or acetate buffer, pH 4.4, containing 0.15 M NaCl).
Sterile, endotoxin-free 10 and 25 ml pipettes, pipetteman tips, microfuge tubes, and 50 ml conical centrifuge tubes (such as Corning 430290) (Fisher Scientific).[3]
Centrifuge with a swinging bucket rotor and adaptors to spin FACs and 50 ml tubes.
5 ml polystyrene round bottom tubes 12 × 75 mm (Falcon 352052, Fisher Scientific).[3]
Microfuge with refrigeration.
Flow cytometer (BD Biosciences).[4]

3a. Procedures

1. Culture bacteria to desired phase of growth (e.g., logarithmic phase) and wash once in DPBS. Resuspend at 1×10^9/ml in DPBS.
2. Add FITC solution such that the final concentration is 0.75 µg/ml, vortex and incubate sample at 37 °C for 20 min (with gentle agitation every 5 min).
3. Wash FITC labeled bacteria twice in DPBS and resuspend at 1×10^8/ml in RPMI-H.
4. Optional: if bacteria need to be opsonized this can be done either before or after labeling with

FITC. To opsonize bacteria, resuspend at 1 × 10^9/ml in RPMI/H containing 25–50% autologous serum and incubate for 30 min at 37 °C with mild agitation.

5. Check efficiency of FITC labeling (530 emission FL-1 channel) by flow cytometry using unlabeled bacteria and FITC labeled sample. Fully compensate sample using FL-2% FL-1. The compensation setting is typically 18–28, depending on the efficiency of FITC labeling. More incorporated label requires increased compensation. Quench the sample with an equal volume of Trypan Blue solution. There should be a clear decrease in fluorescence after quenching (Figure 1).
6. Resuspend isolated PMNs (Support Protocol 1) at 1 × 10^7/ml and place on ice.
7. Coat wells of a 96-well microtiter plate with 20–100% normal human serum (prepared as in Support Protocol 2) for at least one hour and then wash twice with DPBS before adding PMNs. This step helps prevent PMNs from becoming activated by adherence to plastic.
8. Add PMNs at 10^6/100 μl to wells of a 96-well microtiter plate (on ice).
9. Add 10^7/100 μl labeled bacteria to each well.
10. Centrifuge at 400 ×*g* for 7 min at 4 °C.
11. During the centrifugation step (#11) adjust Gains and Threshold on the flow cytometer for PMN analysis (Figure 2). If purified PMNs are used for the assay, there will be a single population of cells at the indicated location with proper instrument settings (Figure 2).
12. Place samples that represent the start of the assay (i.e., time = 0 min) on ice. Place the remaining time points in a 37 °C incubator with 5% CO_2.
13. At the end of the designated time, transfer samples from the 96 well microplate to flow cytometer tubes (Falcon 3755)[3] and place on ice. Of note, it should be possible to modify the assay for use with the Multiwell Autosampler that is manufactured by BD Biosciences for use with their bench-top flow cytometers (e.g., FACSCalibur). Round bottom microplates must be used in place of those with flat-bottoms and a fixation step would be necessary, since the Multiwell Autosampler operates at ambient temperature. However, use of the Multiwell Autosampler would eliminate the need to transfer samples to individual tubes and would increase the throughput of the assay significantly.
14. Read the first sample on the flow cytometer to determine the percent of PMNs associated with bacteria.
15. Add an equal volume of Trypan Blue solution to the same tube and re-read the sample on the flow cytometer. The observed fluorescence represents percent of PMNs with internalized bacteria.
16. After samples are acquired, analyze each sample using a gate around the PMNs to exclude debris and free bacteria (e.g., with BD Bioscience's *Cell Quest*™ software) (Figure 2). The data will be expressed as the percent of cells with associated or internalized bacteria (after adding Trypan Blue).

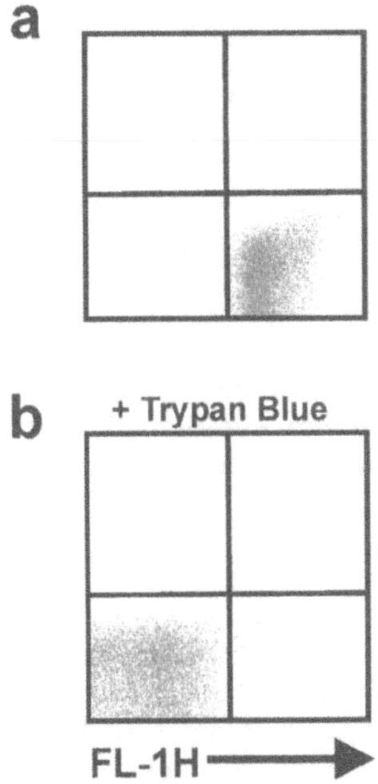

Figure 1. Quenching of FITC-labeled Bacteria Using Trypan Blue. (a) Log phase *Staphylococcus aureus* labeled with FITC. (b) Same sample after quenching with an equal volume of 2 mg/ml Trypan Blue solution (Protocol 1).

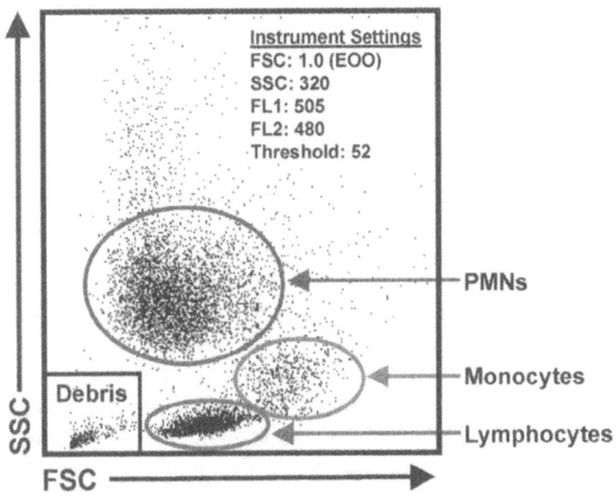

Figure 2. Flow cytometric analysis of leukocyte populations in human whole blood. Figure demonstrates proper FSC and SSC settings for identifying major leukocyte populations. Settings for typical PMN acquisition are inset.

4a. Results and discussion

This assay is most useful for quantitating percent adherence and phagocytosis of bacteria by human PMNs. If set up correctly, results obtained from the assay are quite reproducible and the assay is relatively easy to perform. A typical assay measuring adherence and phagocytosis of Group A *Streptococcus* by human PMNs is shown in Figure 3. The assay can be set up to measure the kinetics of phagocytosis and can be used to determine the optimal time to maximum phagocytosis or onset of phagocytosis. The procedure can be modified to determine the effects of pathogen-derived factors on PMN adherence and phagocytosis.

The most common problem associated with this technique is over-labeling of bacteria with FITC. Although guidelines for labeling and quenching conditions are given above, optimal conditions will have to be determined empirically for each microorganism. Bacteria labeled with excess FITC may not quench efficiently, and the result is only a minor decrease in FL-1 fluorescence after adding Trypan Blue. This problem makes interpretation of the data difficult.

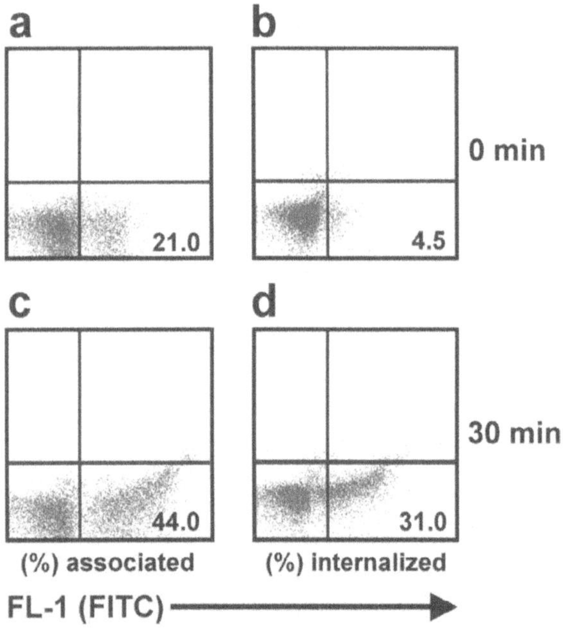

Figure 3. Phagocytosis of Group A *Streptococcus* (GAS) by Human PMNs. GAS was labeled with FITC, mixed with PMNs at a 10:1 bacteria: PMN ratio, and then incubated for the indicated time as described in Protocol 1. Panels indicate PMNs with bound FITC labeled GAS in the absence (a & c) and presence (b & d) of Trypan Blue. Numbers represent the percent of PMNs with bound (associated)/ingested GAS. Instrument settings were optimized for PMNs (Figure 2) and analyzed using a PMN gate with Cell Quest™ software. Compensation for bacterial fluorescence was performed on a sample of FITC labeled-bacteria prior to sample acquisition as in Figure 1. Results show a time-dependent increase in the number of PMNs with both adherent (% associated) and internalized (% internalized) GAS, compare lower right quadrants of panels a & c and b & d.

Another potential difficulty with the assay is determination of the optimal ratio of bacteria to PMNs. The method measures the number of PMNs with internalized bacteria, rather than number of bacteria per PMN. Therefore, the assay should be optimized for phagocytosis of each bacterial species to be tested. An appropriate way to determine an optimal ratio is to set up an experiment with multiple bacteria: PMN ratios and then correlate each sample with microscopy.

An advantage with this assay is that phagocytosis is synchronized, significantly increasing the number of cells with ingested bacteria during peak phagocytosis. Moreover, the kinetics of phagocytosis can be measured accurately. As an alternative, the assay can be performed in microfuge tubes without the centrifugation step and then rotated end over end during the incubation times. However, the number of PMNs containing ingested bacteria at any given time will be significantly less because phagocytosis occurs asynchronously in the PMN population. An asynchronous assay performed in microfuge tubes is useful for determining the effect of exogenously added bacterial factors on PMN phagocytosis because of the continuous mixing of the sample.

5a. Notes on suppliers

1. Sigma-Aldrich, P.O. Box 14508, St. Louis, MO 63178; phone: 1-800-325-3010; http://www.sigma-aldrich.com
2. Invitrogen/Life Technologies Corporation, 9800 Medical Center Drive, P.O. Box 6482, Rockville, MD; phone: 1-800-828-6686; http://www.lifetech.com
3. Fisher Scientific, 2000 Park Lane Drive, Pittsburgh, PA 15275-9943; phone: 1-800-766-7000; http://www.fishersci.com
4. BD Biosciences, Immunocytometry Products, 2350 Qume Drive, San Jose, CA 95131-9936; phone 1-877-232-8995; http://www.bdbiosciences.com

Protocol 2: Measuring PMN phagocytosis and ROS production simultaneously: 4-parameter analysis of host cell response to bacteria

Generation of ROS by PMNs is a vital component of the innate immune response in humans. PMN phagocytosis activates the NADPH (reduced-nicotinomide adenine dinucleotide phosphate) oxidase complex to catalyze the reduction of molecular oxygen to superoxide anion [22]. Superoxide is a short-lived intermediate and a precursor to potent antimicrobial ROS such as hydrogen peroxide, hydroxyl radical and hypochlorous acid. The clearest illustration of the importance of ROS is observed in patients afflicted with chronic granulomatous disease. Individuals with chronic granulomatous disease have a defect in one of the protein components of the NADPH oxidase and suffer from severe and recurrent bacterial infections [11, 18, 22].

Sensitive detection of ROS during phagocytosis can be accomplished by spectrophotometric and/or spectrofluorometric assays [3, 7]. However, these assays are limited to single parameter analysis and multi-parameter analysis is often desired. The protocol and accompanying example shown in Figure 4 describes analysis of PMN-derived ROS and phagocytosis concomitantly (Figure 4). PMNs are loaded with the sodium borohydride-reduced derivative of ethidium bromide, dihydroethidine (DHE), which permeates the cell membrane and is directly oxidized to red-fluorescent ethidium bromide by ROS [29, 30]. The assay is convenient because DHE-labeled PMNs are detected in the FL2 channel (red) and FITC-labeled bacteria are measured in the FL-1 channel (green). Importantly, multiple PMN functions including bacterial adherence, phagocytosis, and ROS production (respiratory burst) can be determined in a single sample. The assay is versatile and this protocol can be adapted to study ROS production concurrently with surface molecule expression.

2b. Materials

Dulbecco's Phosphate Buffered Saline (DPBS) (Sigma D-8662).[1]

Purified human PMNs (Support Protocol 1).

RPMI 1640 medium (Invitrogen/Life Technologies)[2] containing 10 mM HEPES (Sigma H-0887)[1] (RPMI-H).

96 well flat-bottom plates (Fisher Scientific).[3]

Fluorescein isothiocyanate solution (FITC, Sigma F-4274).[1]

Trypan blue solution (Fisher Scientific, J.T. Baker-X242-03)[3], 2 mg/ml in 0.15M NaCl in 0.02 M citrate or acetate buffer pH 4.4).

Dihydroethidine (DHE, Molecular Probes, Inc.).[4]

Sterile, endotoxin-free 10 and 25 ml pipettes, pipetteman tips, microfuge tubes, and 50 ml conical centrifuge tubes (such as Corning 430290) (Fisher Scientific).[3]

Centrifuge with a swinging bucket rotor and adaptors to spin FACs and 50 ml tubes.

Tubes for flow cytometer, such as Falcon 3755 (Fisher Scientific).[3]

Microfuge with refrigeration.

Flow cytometer (BD Biosciences).[5]

3b. Procedures

1. Resuspend PMNs in RPMI-H (25 °C) at 10^7/ml.
2. Add DHE to a final concentration of 25 μM from a 10 mM stock solution made up in DMSO.
3. Incubate samples at room temperature for 20 min in the dark.
4. Immediately following incubation time, transfer samples to ice (keep dark) until ready to use with the phagocytosis assay as described in Protocol 1.
5. Set up flow cytometer for standard 2-color analysis; i.e., adjust instrument settings on each fluorochrome separately to insure proper compensation. Read samples and then analyze with appropriate marker settings (Figure 4a).

4b. Results and discussion

This multiparameter assay quantitates several PMN functions simultaneously. PMN ROS production is measured during concomitant association and/or internalization of bacteria. Importantly, the assay can differentiate ROS production elicited in PMNs by associated bacteria versus those containing ingested pathogens (Figure 4).

Although measuring phagocytosis and ROS production simultaneously is advantageous, there are limitations to flow cytometric analysis of oxidative burst. First, loading of cells with DHE must be determined empirically as overloading significantly increases background fluorescence, making it difficult to detect ROS-derived fluorescence. This is because DHE can slowly oxidize to fluorescent ethidium bromide even in unactivated cells. Additionally, DHE is only one of several possible dyes that can be used to detect ROS. Rothe et al. [29] have shown that DHE is not as sensitive as other fluorometric probes such as dihydrorhodamine 123 (DHR) or 2′7′dicholrofluorescin (DCF). However, a convenient feature of this assay is that it can be easily adapted for use with DCF or DHR, which are detected in the FITC channel (FL-1) [4]. If those dyes are used, the method need only be modified insofar as the bacteria must be labeled with a fluorochrome that can be detected in FL2, 3, or 4 rather than FL1. For example, the cell labeling kit Vybrant™ from Molecular Probes (V-22889)[4] offers lipophilic membrane dyes in three fluorescent channels. These dyes can label most species of bacteria and could be adapted to work with either DCF or DHR.

5b. Notes on suppliers

1. Sigma-Aldrich, P.O. Box 14508, St. Louis, MO 63178; phone: 1-800-325-3010; http://www.sigma-aldrich.com
2. Invitrogen/Life Technologies Corporation, 9800 Medical Center Drive, P.O. Box 6482, Rockville, MD; phone: 1-800-828-6686; http://www.lifetech.com
3. Fisher Scientific, 2000 Park Lane Drive, Pittsburgh, PA 15275-9943; phone: 1-800-766-7000; http://www.fishersci.com

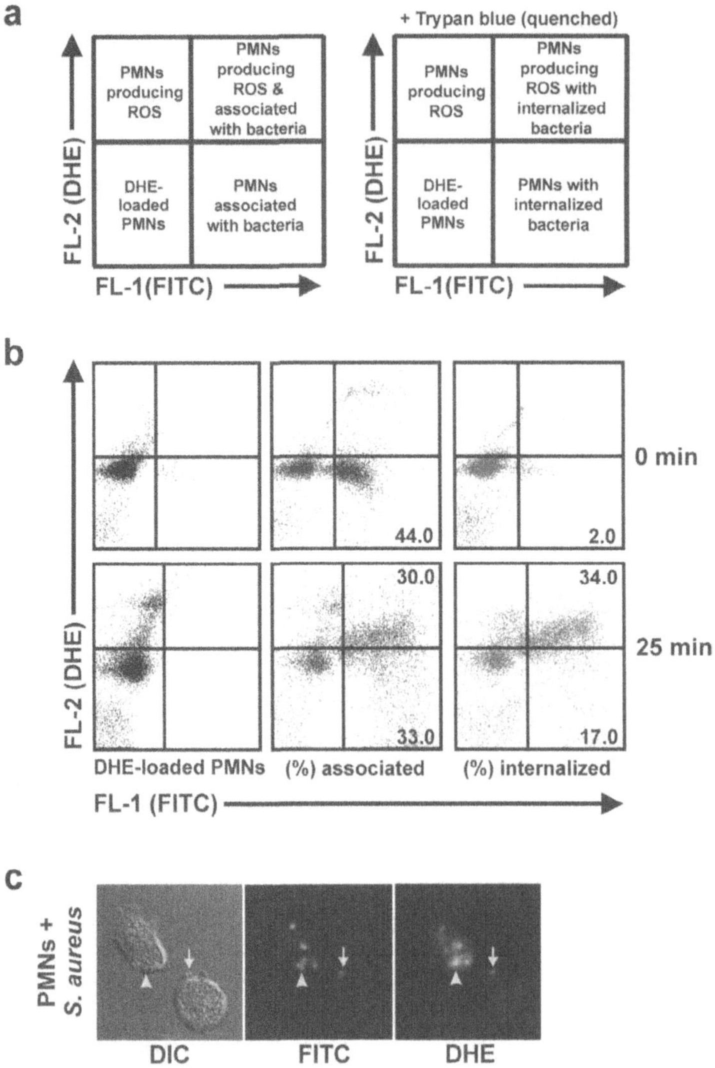

Figure 4. Phagocytosis and ROS production by PMNs exposed to *Staphylococcus aureus* (*S. aureus*). (a) Parameters for each quadrant of the flow cytometer dot plot are defined before and after quenching with Trypan Blue as indicated. (b) Serum opsonized and FITC-labeled *S. aureus* were combined with DHE-loaded PMNs on ice and either analyzed immediately (0 min) or rotated in microfuge tubes at 37 °C for 25 min (25 min) in an asynchronous phagocytosis assay as indicated. (b) Multi-parameter dot plots illustrate PMN phagocytosis of *S. aureus* and concurrent production of ROS. At *t* = 0 min PMNs are unactivated and therefore, do not produce ROS (*t* = 0 min, middle panel, upper left and right quadrants). Binding of *S. aureus* to the PMN surface occurs at 4 °C (*t* = 0 min, middle panel, lower right quadrant) and few have internalized bacteria (*t* = 0 min, right panel, lower right quadrant). At 25 min, the percentage of PMNs with associated and internalized *S. aureus* that are concomitantly producing ROS was similar (*t* = 25 min, compare middle panel, upper right quadrant with right panel, upper right quadrant). Thus, PMNs with ingested bacteria produced ROS. However, due to asynchronous activation, there are populations of PMNs with associated (~33%) or internalized (~17%) bacteria that produce little or no ROS (*t* = 25 min, middle panel, lower right quadrant). The left panels represent DHE-loaded PMNs at both times of incubation. One caveat of the assay is that DHE oxidizes slowly to ethidium bromide over time in unstimulated PMNs and/or in PMNs that become activated during incubation (*t* = 25 min, left panel, upper left quadrant). (c) Microscope-based confirmation of PMN ROS production during phagocytosis of *S. aureus*. FITC-labeled *S. aureus* were combined with DHE-loaded PMNs and incubated at 37 °C for 15 min and then processed for microscopy. Two PMNs with bound or ingested *S. aureus* were imaged for microscopy. The cell indicated with an arrowhead has ingested *S. aureus* and is activated. As a result, DHE has been converted to fluorescent ethidium bromide by PMN-derived ROS. The cell with *S. aureus* indicated by an arrow is in an early stage of phagocytosis (bound) with little concomitant ROS production. Differential interference contrast (DIC, Nomarski (left panel); green-channel fluorescence produced by FITC-labeled *S. aureus* (middle panel); and red-channel fluorescence of ethidium bromide generated from DHE by PMN-derived ROS. PMNs were visualized with an Axiovert 200M inverted fluorescence microscope using a 100X objective (Carl Zeiss, Inc., Thornwood, New York) Micrographs were produced with an Axiocam color digital camera (Carl Zeiss) and images were cropped and annotated in Adobe Photoshop version 6.0 (Adobe Systems Incorporated, San Jose, CA).

4. Molecular Probes, Inc., P.O. Box 22010, Eugene, OR 97402-0469; phone: 1-541-465-8300, http://www.probes.com
5. BD Biosciences, Immunocytometry Products, 2350 Qume Drive, San Jose, CA 95131-9936; phone 1-877-232-8995; http://www.bdbiosciences.com

Protocol 3: Determining phagocytosis-induced PMN apoptosis

PMN apoptosis is important because it prevents damage to healthy tissues that would otherwise occur in the event of necrotic cell lysis. Therefore, apoptosis likely facilitates resolution of inflammation induced by PMN activation. Apoptosis is characterized by several morphological changes including membrane blebbing, cell shrinkage, nuclear condensation and DNA fragmentation [13]. One of the earliest measurable events in cells undergoing apoptosis is redistribution of phosphatidylserine (PS) from the inner leaflet of the plasma membrane to the cell surface (outer leaflet) [15, 36]. Annexin V has high affinity and specificity for PS and therefore, early apoptosis (PS surface expression) can be measured by flow cytometry using annexin V conjugated with FITC or another suitable fluorochrome [36]. Staining of the samples for surface exposed PS in combination with propidium iodide (PI) distinguishes live and dead cells.

Late apoptosis is usually accompanied and/or facilitated by activation of endonuclease and caspase-activated DNAses, which cause extensive DNA cleavage, including breaks at linker regions between nucleosomes. Thus, cells with advanced apoptosis exhibit chromatin or DNA cleavage [40]. This cleavage generates small fragments of double-stranded and/or single stranded DNA with exposed 3-hydroxyl termini that can be detected by modified nucleotides such as bromolated deoxyuridine triphosphates (Br-dUTP). Incorporation of Br-dUTP is catalyzed by terminal deoxynucleotidyl transferase (TdT). This end labeling technique is often called TUNEL for terminal deoxynucleotidyltransferase dUTP nick end labeling, borrowing the nomenclature from the *in situ* nick-end labeling technique [19]. The flow cytometry method is therefore a modified TUNEL assay that does not actually measure nicks, but instead uses deoxynucleotides directly conjugated with fluorochromes to measure end labeling of both single and double-stranded DNA fragments.

As an alternative to detection of apoptosis with TUNEL, late apoptosis can be determined by measuring chromatin condensation with PI. Apoptotic nuclei have increased side-scatter compared with normal nuclei due to the condensation of nuclear chromatin and stain with an accompanying decrease in PI fluorescence [23]. This method provides a simple, repeatable, and cost efficient alternative to TUNEL assays.

The following protocols measure phagocytosis-induced PMN apoptosis and are based on a modification of the basic synchronized phagocytosis assay described above. To measure annexin V staining and TUNEL, samples are processed according to instructions provided with commercial kits. Detection of chromatin condensation by PI [23] is not a kit-based assay and is described in full. An example for each type of flow cytometric measurement is provided. As a rule, it is best to use 2 or 3 of these methods in combination to detect apoptosis and verify that cell death is by apoptosis rather than necrosis. Once apoptosis has been verified by at least two separate methods under the desired assay conditions or with a specific cell type, any of the three procedures can be used independently.

2c. Materials

Dulbecco's Phosphate Buffered Saline (DPBS) (Sigma D-8662).[1]
Purified human PMNs (Support Protocol 1).
RPMI 1640 medium (Invitrogen/Life Technologies)[2] containing 10 mM HEPES (Sigma H-0887)[1] (RPMI-H).
24-well flat-bottom plates.
Opsonized bacteria or suitable particle.
2X hypotonic fluorochrome staining solution; 100 μg/ml PI in 0.2% sodium citrate (Sigma)[1] and 0.2% Triton X-100™ (Sigma T-8532).[1]
Annexin V-FITC Apoptosis Detection Kit II (BD Biosciences, cat. 556570).[4]
Apo-BRDU™ Apoptosis Kit (BD Biosciences, cat. 556405).[4]
Sterile, endotoxin-free 10 and 25 ml pipettes, pipetteman tips, microfuge tubes, and 50 ml conical centrifuge tubes (such as Corning 430290) (Fisher Scientific).[3]
Centrifuge with a swinging bucket rotor and adaptors to spin FACS and 50 ml tubes.
Tubes for flow cytometer, such as Falcon 3755 (Fisher Scientific).[3]
Microfuge with refrigeration.
Flow cytometer (BD Biosciences).[5]

3c. Procedures

For all assays that measure apoptosis, set up the phagocytosis assay from Protocol 1, steps 1–13 with the following modifications. Use 2×10^6 PMNs (double the amount used in Protocol 1) and a correspondingly increased number of bacteria or other suitable particle. The assays should be performed in 24-well plates rather than those containing 96 wells. Also, do not label bacteria with fluorochrome.

Annexin staining and DNA fragmentation (TUNEL) using commercially available kits

1. Follow the manufacturers protocol for annexin-V staining (Figure 5a) using the following modifications for PMNs. Cells should be resuspended at 1×10^6/100 µl in 1.5 ml microfuge tubes and stained with annexin V-FITC and PI before transferring to round bottom polystyrene FACs tubes.
2. Follow the manufacturers protocol for the APO-BRDU™ kit with the following modifications.
   a. Resuspend cells in 100 µl of DPBS in 1.5 ml tubes to reduce fixation volume.
   b. Add 1 ml of 4.0% paraformaldehyde for fixing PMNs instead of the recommended 1.0%.
   c. Transfer cells to FACs tubes for DNA staining and stain for 90 min instead of the recommended 60 min.

Chromatin condensation: detection with propidium iodide

1. At the designated time, transfer PMNs to microfuge tubes and centrifuge at 400 ×*g* for 5 minutes (4 °C). Use non-apoptotic cells as a control for PI staining.
2. Resuspend cells in 100 µl of 2X hypotonic fluorochrome staining solution and incubate overnight in the dark at 4 °C.
3. Analyze by flow cytometry using non-apoptotic cells as a reference for setting up fluorescence. Non-apoptotic cells will stain uniformly and are characterized by a bright and narrow peak in the FL-2 channel. The non-apoptotic control should be set up so this peak occurs at approximately 10^3 in the FL-2 channel. Apoptotic cells are characterized by reduced FL-2 fluorescence and a broadened peak (Figure 5b). Debris normally fluoresces at approximately 10^1 or less in the FL-2 channel and should be removed from the analysis by setting a specific gate.

4c. Results and discussion

The following general conclusions can be made when determining apoptosis by PS detection. Cells that stain positive for annexin V only are undergoing early apoptosis, cells that stain with PI only are dead, and cells that stain positive for both annexin V and PI are either in late apoptosis or in early stages of necrosis (Figure 5a). To differentiate apoptosis and necrosis, annexin V staining can be coupled with an assay that measures either chromatin condensation or DNA fragmentation. False positives can occur in this assay, therefore, it is advisable to use an additional control preincubated with unconjugated annexin V followed by incubation with annexin V-FITC to determine specificity of binding.

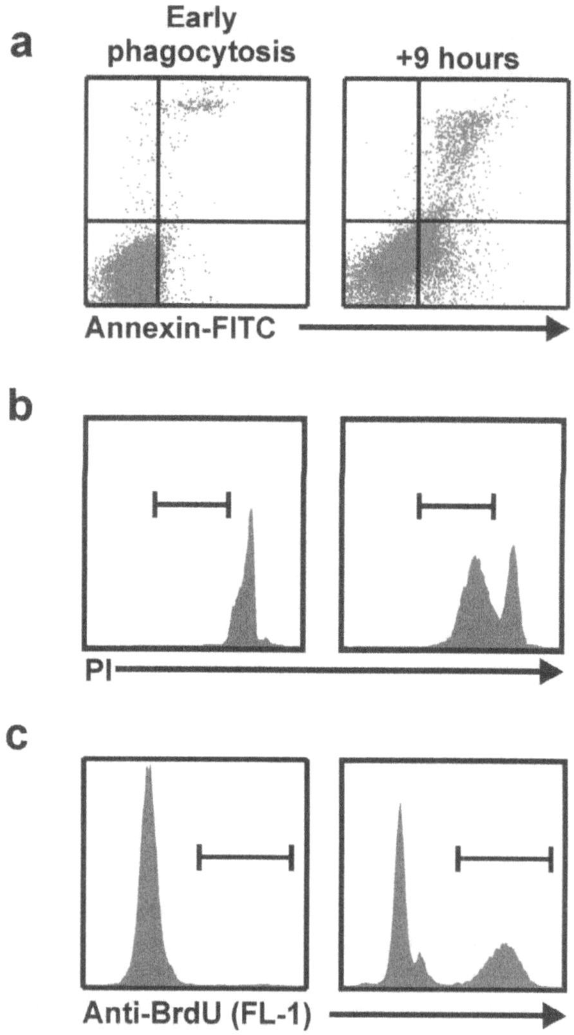

Figure 5. Phagocytosis-induced apoptosis in human PMNs. Synchronized phagocytosis was performed as described in Protocol 1 and samples were processed for detection of apoptosis (Protocol 3). (a) PMNs were stained with both annexin V coupled to FITC, and PI. Lower right quadrant represents percentage of PMNs with surfaced expressed PS. Cells stained with both annexin V and PI (upper right quadrant) are either necrotic or in very late stages of apoptosis. (b) PMN nuclei were stained with PI to detect chromatin condensation. Note the increase in the number of apoptotic cells over time corresponding to a decrease in PI fluorescence. (c) PMN apoptosis was detected with TUNEL using an anti-BrdU monoclonal antibody conjugated with FITC. The marker in (b) and (c) indicates apoptotic cells.

Measuring PMN apoptosis with the modified TUNEL (Apo-BRDU) assay is very reproducible once the fixation procedure is optimized for the cell type. However, even with proper fixation, clumping of PMNs at late time points can occur. It is important to disperse these clumps before proceeding with the DNA labeling reaction to insure equivalent staining. Also, if DNA cell cycle information is desired, follow the manufacturers protocol for staining with PI/RNase A. An example of PMN apoptosis determined with the modified TUNEL is shown in Figure 5c.

The method described for measuring chromatin condensation with PI is fast, simple, inexpensive and very reproducible. If problems occur with the staining, e.g. the control cells do not elicit a single narrow peak, it is most likely due to improper preparation of the hypotonic solution.

5c. Notes on suppliers

1. Sigma-Aldrich, P.O. Box 14508, St. Louis, MO 63178; phone: 1-800-325-3010; http://www.sigma-aldrich.com
2. Invitrogen/Life Technologies Corporation, 9800 Medical Center Drive, P.O. Box 6482, Rockville, MD; phone: 1-800-828-6686; http://www.lifetech.com
3. Fisher Scientific, 2000 Park Lane Drive, Pittsburgh, PA 15275-9943; phone: 1-800-766-7000; http://www.fishersci.com
4. BD Biosciences, Immunocytometry Products, 2350 Qume Drive, San Jose, CA 95131-9936; phone 1-877-232-8995; http://www.bdbiosciences.com

Support Protocol 1: isolation of human polymorphonuclear leukocytes (PMNs)

This procedure is a modification of the original method described by Boyum [5], and yields preparations that contain >99% PMNs. Peripheral blood mononuclear cells (PBMCs) can be isolated with the same procedure.

2d. Materials

Sterile, non-pyrogenic 3.0% dextran, 0.9% NaCl solution. Prepare by mixing 30 g dextran T-500.
(Amersham-Pharmacia Biotech, cat. 17-0320-02)[1] with 1 L 0.9% sodium chloride (Irrigation USP, Baxter Healthcare, cat. 2F7124).[2] Filter sterilize using a 0.22 μM filter when dissolved.
Sterile, non-pyrogenic 0.9% sodium chloride (Irrigation USP, Baxter Healthcare, cat. 2F7124).[2]
Sterile, non-pyrogenic 1.7% sodium chloride solution. Prepare with RNase free NaCl (Sigma)[3] and irrigation-grade water (Irrigation USP, Baxter Healthcare, cat. 2F7114).[2]
Ficoll-Paque™ PLUS (1.077 g/L, Amersham Biosciences, cat. 17-1440-03).[1]
Sterile, non-pyrogenic water (Irrigation USP, Baxter Healthcare, cat. 2F7114).[2]
Sterile, endotoxin-free 10 and 25 ml pipettes, pipetteman tips, microfuge tubes, and 50 ml conical centrifuge tubes (such as Corning 430290) (Fisher Scientific).[5]
Centrifuge with a swinging bucket rotor and adaptors to spin 50 ml tubes.
RPMI 1640 medium (containing 10 mM HEPES. Prepare by adding 5 ml sterile, non-pyrogenic HEPES (Sigma, cat. H-0887)[3] to 500 ml RPMI 1640 medium (Invitrogen/Life Technologies).[4]
*Heparinized, venous human blood, 20–200 ml.
2.0% acetic acid solution. Prepare by mixing 200 μl glacial acetic acid with 9.8 ml water.
Sterile, endotoxin-free pipettes, 10 and 25 ml, 50 ml centrifuge tubes, pipetteman tips (Fisher Scientific).[5]
Centrifuge with a swinging bucket rotor and adaptors to spin 50 ml tubes.
Hemocytometer and light microscope.

*NOTE: human blood samples can only be obtained in accordance with a Human Subjects Protocol that has been approved by an Institutional Review Board for Human Subjects. Venipuncture must be performed by a trained and competent phlebotomist in accordance with the guidelines set forth in the Human Subjects Protocol. Human blood and blood cells are biohazards and appropriate precautions should be implemented during handling.

3d. Procedures

1. Combine whole blood with an equal volume of 3.0% Dextran/0.9% NaCl (use screw-cap 50 ml conical tubes such as Corning 430290). Mix blood gently by inverting the tube and let stand at room temperature for 20–25 min.
2. During the sedimentation, prepare four 50 ml tubes containing the following solutions: 35 ml 0.9% NaCl, 20 ml 1.7% NaCl, 20 ml dH_2O, and a tube with 11–12 ml Hypaque-Ficoll™.
3. After 20 min of dextran sedimentation, transfer top (leukocyte-rich) layer from dextran-blood mixture to a fresh tube.
4. Centrifuge at 650 $\times g$ for 10 min to pellet all cells.
5. Aspirate supernatant, and discard into a container with 10% bleach.
6. Resuspend the pellet in 10 ml 0.9% NaCl and then add the remaining 25 ml of 0.9% NaCl.
7. Carefully underlay 10 ml of Hypaque-Ficoll™ (HF) (beneath the cell suspension) using a hand pipette.
8. Spin at 350–400 $\times g$, for 25 min, no brake.
9. After the spin, there should be two discernable cell populations. Peripheral blood mononuclear cells (PBMCs) sediment to the top of the HF and should be aspirated carefully from the NaCl-HF interface. PMNs and erythrocytes sediment through the HF and form the pellet at the bottom of the tube. After removing PBMCs, carefully aspirate all of the remaining liquid in the tube. Wipe the inside of the tube with sterile cotton swabs to remove cells adhering to the side.

10. Lyse erythrocytes in 20 ml H_2O by pipetting for 15–20 s. Note: it is critical that this step be performed consistently and rapidly and that 1.7 ml NaCl is added immediately (step 11). This step can result in pre-activated, damaged, or lysed PMNs if done incorrectly.
11. Immediately add 20 ml of 1.7% NaCl to the cells and mix by inverting the tube several times.
12. Spin at 350–400 ×*g* for 8 min.
13. Discard supernatant containing lysed erythrocytes and resuspend PMNs in 5–10 ml of RPMI/H and enumerate by microscopy. Dilute PMNs 1:20 (10 μl into 190 μl 2% acetic acid) and count with a hemocytometer. PMNs can be distinguished from any contaminating PBMCs using this counting method.
14. For any of the synchronized phagocytosis assays, the cells should be stored on ice prior to use. It is critical that repeated warming and chilling of the cells be avoided and cells should be used within 4 hours after isolation.

4d. Results and discussion

If carefully done, the procedure yields PMNs which are >99% pure. Most of the cells will be neutrophils (~95–98%) and the rest predominantly eosinophils (~2–5%). Basophils typically comprise less than 1% of the PMN preparation. It is crucial that all reagents contain little (~<10–25 pg/ml) or no endotoxin, since PMN function will be significantly altered upon exposure to lipopolysaccharide. The cells should not be chilled and warmed repeatedly, as that will prime and/or activate PMNs. Therefore, it is best to perform the procedure at ambient temperature and then chill cells on ice only once for synchronization of the phagocytosis assay. Another critical step is hypotonic lysis of erythrocytes. This step must be performed efficiently or cells will activate and/or become damaged. If PMNs clump after the preparation, the problem can usually be attributed to the hypotonic lysis step. It usually takes an inexperienced individual several attempts to purify cells that are unactivated and thus, usable in the assays mentioned above.

5d. Notes on suppliers

1. Amersham-Pharmacia Biotech, 800 Centennial Avenue, P.O. Box 1327, Piscataway, NJ 08855-1327; phone: 1-800-526-3593; http://www.apbiotech.com
2. Baxter Healthcare Corporation, One Baxter Parkway DF3-2E, Deerfield, IL 60015; phone: 1-847-948-4675
3. Sigma-Aldrich, P.O. Box 14508, St. Louis, MO 63178; phone: 1-800-325-3010; http://www.sigma-aldrich.com
4. Invitrogen/Life Technologies Corporation, 9800 Medical Center Drive, P.O. Box 6482, Rockville, MD; phone: 1-800-828-6686; http://www.lifetech.com
5. Fisher Scientific, 2000 Park Lane Drive, Pittsburgh, PA 15275-9943; phone: 1-800-766-7000; http://www.fishersci.com

Support Protocol 2: isolation of serum from whole blood

2e. Materials

Dulbecco's Phosphate Buffered Saline (DPBS) (Sigma D-8662).[1]
*Unheparized human blood (see *NOTE, above for working with human blood)
30 ml sterile glass Corex tube (Fisher Scientific).[2]

3e. Procedures

1. Transfer 10–20 ml unheparinized blood into a sterile Corex high-speed glass centrifuge tube (Fisher Scientific).[1]
2. Incubate blood at 37 °C at least 30 min.
3. Spin sample at 3500 ×*g* for 10–15 min at 4 °C.
4. Aspirate supernatant (serum) and use immediately or store at –80 °C. For efficient opsonization of particles or bacteria, serum should be thawed only once. For coating of microplates and tissue culture dishes, dilute serum in sterile DPBS.

4e. Results and discussion

The serum should be dull to bright yellow with little or no reddish tinge. A red or pink hue indicates a significant amount of contaminating erythrocytes and these should be removed by further centrifugation (3500 ×*g*, for 5 min in a microfuge). If, after further clarification the red color persists, it is likely that some degree of hemolysis occurred during the phlebotomy. The serum should be filter sterilized immediately if it is to be used for *in vitro* culture of leukocytes (e.g., PBMCs).

5e. Notes on suppliers

1. Sigma-Aldrich, P.O. Box 14508, St. Louis, MO 63178; phone: 1-800-325-3010; http://www.sigma-aldrich.com
2. Fisher Scientific, 2000 Park Lane Drive, Pittsburgh, PA 15275-9943; phone: 1-800-766-7000; http://www.fishersci.com

References

1. Ahmed HJ, Johansson C, Svensson LA, Ahlman K, Verdrengh M, Lagergard T (2002). *In vitro* and *in vivo* interactions of haemophilus ducreyi with host phagocytes. Infect Immun 70: 899–908.

2. Allen LA, DeLeo FR, Gallois A, Toyoshima S, Suzuki K, Nauseef WM (1999). Transient association of the nicotinamide adenine dinucleotide phosphate oxidase subunits p47phox and p67phox with phagosomes in neutrophils from patients with X-linked chronic granulomatous disease. Blood 93: 3521–3530.
3. Babior BM, Kipnes RS, Curnutte JT (1973). Biological defense mechanisms. The production by leukocytes of superoxide, a potential bactericidal agent. J Clin Invest 52: 741–744
4. Bassoe CF, Smith I, Sornes S, Halstensen A, Lehmann AK (2000). Concurrent measurement of antigen- and antibody-dependent oxidative burst and phagocytosis in monocytes and neutrophils. J Immunol Methods 21: 203–220.
5. Boyum A (1968). Isolation of mononuclear cells and granulocytes from human blood. Isolation of mononuclear cells by one centrifugation, and of granulocytes by combining centrifugation and sedimentation at 1 g. Scand J Clin Lab Invest Suppl 97: 77–89.
6. Cunningham MW (2000). Pathogenesis of group a streptococcus infections. Clin Microbiol Rev 3: 470–511.
7. DeLeo FD, Allen LA, Apicella MA, Nauseef WM (1999). NADPH oxidase activation and assembly during phagocytosis. J Immunol 163: 6733–6740.
8. Ernst JD (2000). Bacterial inhibition of phagocytosis. Cell Microbiol 2: 379–386.
9. Gao LY, Kwaik YA (2000). The modulation of host cell apoptosis by intracellular bacterial pathogens. Trends Microbiol 8: 306–313.
10. Gao LY, Kwaik YA (2000). Hijacking of apoptotic pathways by bacterial pathogens. Microbes Infect 2: 1705–1719.
11. Holland SM, Gallin JI (1998). Evaluation of the patient with recurrent bacterial infections. Annu Rev Med 49: 185–199.
12. Jansen WTM, Gootjes J, Zelle M, Madore DV, Verhoef J, Snippe H, Verheul AFM (1998). Use of highly encapsulated streptococcus pneumoniae strains in a flow-cytometric assay for assessment of the phagocytic capacity of serotype-specific antibodies. Clin Diagn Lab Immunol 5: 703–710.
13. Kerr JF, Wyllie AH, Currie AR (1972). Apoptosis: a basic biological phenomenon with wide-ranging implications in tissue kinetics. Br J Cancer 26: 239–257.
14. Kobayashi SD, Voyich JM, Buhl CL, Stahl RM, DeLeo FR (2002). Global changes in gene expression by human polymorphonuclear leukocytes during receptor-mediated phagocytosis: cell fate is regulated at the level of gene expression. Proc Natl Acad Sci USA 99: 601–6906.
15. Koopman G, Reutelingsperger CP, Kuijten GA, Keehnen RM, Pals ST, van Oers MH (1994). Annexin V for flow cytometric detection of phosphatidylserine expression on B cells undergoing apoptosis. Blood 84: 1415–1420.
16. Lehmann AK, Halstensen A, Bassoe CF (1998). Flow cytometric quantitation of human opsonin-dependent phagocytosis and oxidative burst responses to meningococcal antigens. Cytometry 33: 406–413.
17. Lehmann AK, Sornes S, Halstensen A (2000). Phagocytosis: measurement by flow cytometry. J Immunol Methods 243: 229–242.
18. Lekstrom-Himes JA, Gallin JI (2000). Immunodeficiency diseases caused by defects in phagocytes. Adv Immunol 343: 1703–1714.
19. Li X, Darzynkiewicz Z (1995). Labeling dna strand breaks with brd utp. Detection of apoptosis and cell proliferation. Cell Prolif 28: 571–579.
20. Martinez JE, Romero-Steiner S, Pilishvili T, Barnard S, Schinsky J, Goldblatt D, Carlone DM (1999). A flow cytometric opsonophagocytic assay for measurement of functional antibodies elicited after vaccination with the 23-valent pneumococcal polysaccharide vaccine. Clin Diagn Lab Immunol 6: 581–586.
21. McCloskey PS, Salo RJ (2000). Flow cytometric analysis of group b streptococci phagocytosis and oxidative burst in human neutrophils and monocytes. FEMS Immunol Med Microbiol 27: 59–65.
22. Nauseef WM, Clark RA (2000). Granulocytic phagocytes. Chapter in: Mandell GL, Bennet JE, Dolin R. (eds), Basic principles in the diagnosis and management of infectious disease, 5th ed. New York: Churchill Livingstone, pp 89–112.
23. Nicoletti I, Migliorati G, Paglizcci MC, Grignani F, Riccardi C (1991). A rapid and simple method for measuring thymocyte apoptosis by propidium iodide staining and flow cytometry. J Immunol Methods 139: 271–279.
24. Park MA, Lee MJ, Lee SH, Jung DK, Kwak JY (2002). Anti apoptotic role of phopholipase d in spontaneous and delayed apoptosis of human neutrophils. FEBS Lett 519: 45–49.
25. Perskvist N, Long M, Stendahl O, Zheng L (2002). Mycobacterium tuberculosis promotes apoptosis in human neutrophils by activating caspase-3 and altering expression of bax/bcl-xl via an oxygen dependent pathway. J Immunol 168: 6358–6365.
26. Plested JS, Berne FL, Couli PA, Makepeace K, Lehmann AK, Mackinnon FG, Griffiths HG, Herbert MA, Richards JC, Moxon ER (2001). Functional opsonic activity of human serum antibodies to inner core lipopolysaccharide (gale) of serogroup b meningoccoci measured by flow cytometry. Infect Immun 69: 3203–3213.
27. Rodriguez ME, van der Pol WL, van de Winkel JG (2001). Flow cytometry-based phagocytosis assay for sensitive detection of opsonic activity of pneumococcal capsular antibodies in human sera. J Immunol Methods 252: 33–44.
28. Rozenberg-Arska M, Salters M, van Strijp J, Geuze J, Verhoef J (1985). Electron microscopic study of phagocytosis of escherichia coli by human polymorphonuclear leukocytes. Infect Immun 50: 852–859.
29. Rothe G, Valet G (1990). Flow cytometric analysis of respiratory burst activity in phagocytes with hydroethidine and 2′7′-dichlorofluorescin. J Leuk Biol 47: 440–448.
30. Rothe G, Valet G (1994). Flow cytometric assays of oxidative burst activity in phagocytes. Methods Enzymol 233: 539–548.
31. Savill J (1997). Apoptosis in the resolution of inflammation. J Leuk Biol 61: 375–380.
32. Savill JS, Wyllie AH, Henson JE, Walport MJ,

Henson PM, Haslett C (1989). Macrophage phagocytosis of aging neutrophils in inflammation. Programmed cell death in the neutrophil leads to its recognition by macrophages. J Clin Invest 83: 865–875.

33. Stendahl O, Hed J, Kihlstrom E, Magnusson KE, Tagesson C (1977). Phagocytic internalization and the requirement for membrane perturbation. FEMS Lett 81: 118–120.
34. Underhill DM, Ozinsky A (2002). Phagocytosis of microbes: complexity in action. Annu Rev Immunol 20: 825–852.
35. van Eeden S, Klut M, Walker B, Hogg J (1999). The use of flow cytometry to measure neutrophil function. J Immunol Methods 232: 23–43.
36. Vermes I, Haanen C, Steffens-Nakken H, Reutelingsperger C (1995). A novel assay for apoptosis. Flow cytometric detection of phosphatidylserine expression on early apoptotic cells using fluorescein labeled annexin V. J Immunol Methods 184: 39–51.
37. Villasenor-Sierra A, McShan WM, Salmi D, Kaplan EL, Johnson DR, Stevens DL (1999). Variable susceptibility to opsonophagocytosis of group a streptococcus m-1 strains by human immune sera. J Infect Dis 180: 1921–1928.
38. Weinrauch Y, Zychlinsky A (1999). The induction of apoptosis by bacterial pathogens. Annu Rev Microbiol 53: 155–187.
39. Whyte MK, Meagher LC, MacDermot J, Haslett C (1993). Impairment of function in aging neutrophils is associated with apoptosis. J Immunol 150: 5124–5134.
40. Wyllie HH, Morris RG, Smith AL, Dunlop D (1984). Chromatin cleavage in apoptosis: association with condensed chromatin morphology and dependence on macro-molecular synthesis. J Pathol 142: 67–77.

Author for correspondence: Frank R. DeLeo, Ph.D., Rocky Mountain Laboratories, National Institute of Allergy and Infectious Diseases, National Institutes of Health, 903 South 4th Street, Hamilton, MT 59840, USA
Phone: 406-363-9448; Fax: 406-363-9336
E-mail: fdeleo@niaid.nih.gov

Methods in Cell Science **24**: 91–97 (2002).

Flow cytometric techniques for the detection of microorganisms

Hazel M. Davey
Institute of Biological Sciences, University of Wales, Aberystwyth, Ceredigion, SY23 3DD Wales, UK

Abstract. Flow cytometry (FCM) is a technique, which allows one to analyse cells rapidly and individually, and permits the quantitative analysis of distributions of a property or properties in a population. It therefore offers many advantages over conventional measurements for the analysis of biological cells. Historically the technique has been widely applied for the study of mammalian cells, but its use in microbiology has been more limited; this is mainly a consequence of the smaller size of microbes, which results in the smaller optical signals that can be obtained from them. Developments in light sources and optics, together with brighter, spectrally-diverse dyes have reduced this barrier over recent years and the flow cytometer is now an essential tool in many microbiological research establishments. FCM has an increasing role to play in the detection of microbes in both industrial and clinical settings. Environmental monitoring to prevent outbreaks of human diseases such as cryptosporidiosis and Legionnaires' disease and to detect acts of biowarfare or bioterrorism are all amenable to flow cytometric study. This review seeks to highlight the role of the flow cytometer in the detection of microbial cells.

Key words: Bacteria, Detection, Flow cytometry, Microbiology, Yeast

Abbreviations: FACS = Fluorescence-activated cell sorting; FCM = Flow cytometry; FITC = Fluorescein isothiocyanate

1. Introduction

Flow cytometry (FCM) is a technique that permits the rapid, optical analysis of individual cells, and has many advantages over conventional cytometry. Firstly, since measurement rates of 1000 to 25,000 cells.sec^{-1} can be achieved (depending on the make/model of the instrument), flow cytometric data sets often represent 10,000 to 500,000 cells for a given population. This leads to the relatively straightforward acquisition of statistically significant results (cf. measurements made by light or fluorescence microscopy on at best a few hundred cells). By analysing many more cells than would be possible by conventional cytometry, rare cell types are more likely to be detected. Secondly, since FCM uses very sensitive electronic detectors to measure the intensity of scattered light or fluorescence at a given wavelength, different intensities of light scatter/fluorescence can be distinguished. By calibrating an instrument with samples of known size or fluorescent intensity, it is possible to obtain quantitative measurements of sample heterogeneity. Thirdly, by using dichroic filters to optically separate light of different wavelength, flow cytometric measurements can be made on several different characteristics of each cell. Such multiparametric measurements are useful because they allow one to correlate the different characteristics and thus define subpopulations and/or distinguish between different cell types. Typical commercial flow cytometers allow 5–10 different parameters (e.g. size, protein content, DNA content, lipid content, antigenic properties, enzyme activity etc) to be collected for each cell, allowing a multidimensional representation of a population to be obtained. Finally, since measurements are made on single cells, heterogeneity within the population can be detected and quantified. Thus, flow cytometric analysis enables the investigator to build up a multidimensional representation of the individual cells within a population.

Whilst all commercial flow cytometers have the advantages described above, some specialised instruments (cell sorters) are able to physically separate cells on the basis of user-defined characteristics. Cells may be bulk-sorted for biochemical analysis in order to gain an understanding of how different cohorts of cells vary with respect to a character of interest [1]. Alternatively, individual cells may be sorted onto microscope slides for confirmation of their identities [2]. A powerful approach is to sort single cells into microtitre plates or onto agar plates for recovery [3, 4]. Providing that appropriate cell staining and sample preparation methods have been used that maintain viability, sorted cells can be grown to give clonal colonies or broth suspensions. This has particular advantages in molecular biology for isolating mutants [5], and is also of use in strain improvement programmes in industrial microbiology [6, 7].

Over recent years a number of reviews of microbial FCM have been published [see e.g. 8, 9, 10]. The purpose of this review is to highlight the use of FCM for the detection of microorganisms.

2. Detection of microorganisms

There is a continuing need to detect (and in some cases quantify and identify) microbial cells in a variety of environments. In the research laboratory this may be the relatively simple task of monitoring growth of an axenic culture in a non-particulate medium. However, in industrial or clinical labs and particularly with environmental samples, it is often necessary to enumerate mixed populations of bacteria and/or yeasts against a background of other biological and non-biological particulates.

3. FCM for counting microorganisms in the research laboratory

In the microbial research laboratory, it is often necessary to determine the number of (viable) cells in a sample or culture of interest. This is usually achieved by plating the sample (diluted as required) on to an agar plate. There are several problems associated with this technique, the greatest of which is the length of time required to obtain the results. For some slowly-growing organisms it may take in excess of a week to determine how many cells were 'viable' in the original sample, and even when the sample contains quickly-growing organisms and the plates are incubated under optimal growth conditions a minimum of overnight growth is usually required before the resulting colonies can be counted. In many situations this may be too long to be of use and consequently alternatives to plate counts have been proposed in order to decrease the time required to determine numbers of viable cells [see e.g. 11]. Some flow cytometers permit direct counting by determining the number of particles in a known volume of sample. In other systems it is necessary to 'spike' the sample with an easily-distinguishable bead or control cell type of known concentration thus allowing ratiometric counting [12]. By incubating cells with a suitable fluorescent dye [4] whose uptake or exclusion depends on viability of the cell, a total count and a viable count can be obtained simultaneously (Figure 1).

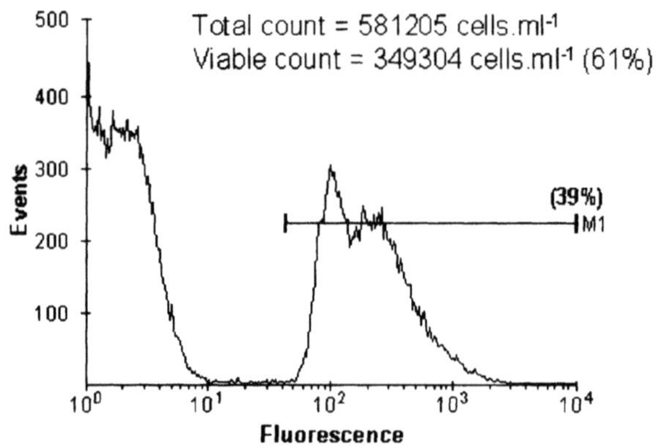

Figure 1. Flow cytometric analysis of a mixture of live and dead (ethanol fixed) *Saccharomyces cerevisiae* stained with 5 μM Propidium Iodide. Flow cytometry was performed on a Partec PAS III instrument using 488 nm excitation and forward scatter was used as a trigger. Fluorescence was collected *via* a 630 nm band-pass filter. The intact membrane of live cells excludes PI; in contrast PI freely enters dead (permeabilised) cells and its fluorescence is enhanced on binding to nucleic acids. Thus dead cells will fluoresce more brightly than live cells. Total counts were performed directly by the instrument by counting the total number of events in a fixed volume (~200 μl). The viable count was determined from the number of events outside of the region M1 which had been set to include only dead yeast during the analysis of control samples.

4. Industrial applications of microbial detection by FCM

4.1. *Quality control*

Since FCM involves the study of individual particles it can be used to detect bacteria/yeasts against a background of other particulates. Thus FCM is emerging as a useful tool in the Quality Control environment [13]. Several early microbial FCM papers concentrated on the use of FCM for the detection of bacteria (generally), or specific species of bacteria against the complex particulate matrix of food samples. This led to the development of instruments designed specifically for this purpose [14]. In a more recent study Gunasekera et al. [15] demonstrated the application of FCM to microbial analysis of milk. Normally such assays would be hampered by the presence of milk proteins and lipid particles. However, following use of an enzymatic clearing method, the detection limit of the FCM assay was found to be less than or equal to 10^4 bacteria.ml of milk^{-1}.

In pharmaceutical laboratories there is a need to control the microbiological quality of water, raw materials and products. FCM can be used to determine the total microbial content in pharmaceutical samples. The use of rapid rather than culture-based methods leads to near real time monitoring of processes and finished products and allows faster release of products from production facilities [16].

One of the most demanding situations where FCM has been used to detect a given organism must surely be the detection of wild yeast against a background of brewer's yeast [17]. Contamination of the brewing process can produce noticeable spoilage effects even when wild yeasts are present at a concentration of 1 wild yeast per 10^4 culture yeast. Jespersen and colleagues showed that by careful formulation of a growth medium that favoured the growth of the wild yeasts followed by flow cytometric analysis, they could detect 1 wild yeast per 10^6 culture yeast within 24–72 hours.

Whilst the examples above illustrate the role that FCM has to play in the detection of microbial *contamination* of industrial processes and products, it is useful also to consider the value of the flow cytometric approach to the monitoring of industrial fermentations. For example dried yeast is used as an inoculum for many production processes in the baking, brewing and wine-making industries. The viability and fermentative activity (vitality) of the yeast used in these processes is a critical parameter in the quality of the finished product. FCM together with appropriate fluorescent stains can be used to predict the abilities of populations of dried bakers' and wine yeasts to ferment after rehydration. Recently Attfield and colleagues [18] demonstrated that fluorescent viability stains combined with FCM provide a rapid and accurate method for determining the quality of a yeast inoculum. A good relationship was found between the fermentative efficiency of yeast cells and their ability to exclude fluorescent stains such as propidium iodide or oxonol.

Heterogeneity in the physiological status of microorganisms may also have profound consequences on the outcome of industrial fermentations. Traditional culture-based techniques often fail to detect stressed, sub-lethally injured, non-culturable (under the conditions provided) or dead cells. However these cells can have an influence on the outcome of fermentations and their presence may contribute to the problems of industrial 'scale-up'. By using fluorescent stains with different modes of action with respect to viability determination (e.g. membrane intactness, membrane energisation, presence of functioning extrusion pumps etc) it has been demonstrated that FCM can resolve and enumerate sub-populations with differing physiological properties [19]. This 'knowledge' can then be used to formulate more accurate mathematical models for the prediction of bioprocess performance.

4.2. *Strain improvement*

Provided that a suitable assay can be devised, fluorescence-activated cell sorting (FACS) can be used to separate single cells or cohorts of cells with desirable characteristics. This approach can be exploited to obtain 'improved strains' for industrial purposes.

An early microbial FCM paper [6] described the use of a sorter specially adapted for the isolation of single cells onto petri dishes or microtitre plates. Such technology has since become available on several commercial machines. Betz and colleagues used FCM to separate a subpopulation of morphologically distinct spores of *Rhizopus arrhizus* identified on the basis of differences in their light scatter signals. From ~8,300 clones, 10 were isolated that produced approximately five-to six-fold more fungal lipase activity, when compared either to the original strain or to a random selection of reisolated clones from the sorted population.

More recently a flow cytometric approach for selection of novel strains of the yeast *Saccharomyces cerevisiae* has been developed. Mating parental strains and screening the resulting hybrids is a common approach to producing strains with desirable industrial traits. However when the parental strains and the hybrids are all capable of growth on the same medium, separating the novel strains from the parental strains can be problematic. With industrial strains, the introduction of genetic markers (such as antibiotic resistance) to aid separation of the hybrids from the parental strains is undesirable because these may affect fermentation performance. Additionally, yeast are often used in food manufacture where it is undesirable to use antibiotics and antibiotic-resistant strains. To overcome these problems Bell and colleagues [7] used a flow cytometric approach based on cell tracking dye technology (dyes which are 'diluted out' by cell division). One parental strain was loaded with CT-Green BODIPY (green) and the second strain was loaded with BR (orange). Following successful mating of the two parental strains a small subpopulation (1%) of the 2 million cells formed a third dual-stained population. Cell sorting and microscopic observation were used to confirm that this dual-stained population contained cell clusters containing at least one cell of each parent. This region was sorted twice (to improve purity), and 50 isolates were obtained. Three of these isolates had PCR fingerprints that were consistent with being the result of hybridization between the two parental strains. The three putative rare mated products had the phenotypes of both parents, including *lacZ* expression, derived from one parent, and prototrophy derived from the other. Thus FACS was shown to be a very effective method for enriching the population for hybrids and yielded a small, viable population that could be further screened by conventional means.

5. Clinical applications of microbial detection by FCM

The detection of bacteria or yeasts in body fluids is important for diagnosis of a variety of diseases. Urinary sediments may contain a variety of particulates including red and white blood cells, epithelial cells, bacteria and inorganic chemical crystals. The presence and concentrations of these particulates is used in the diagnosis of a range of diseases and disorders. Flow cytometers designed specifically for urinalysis are now available commercially that allow the simultaneous determination of concentrations of bacteria together with red and white blood cells, squamous epithelial cells etc [20]. It has been shown that such devices are more sensitive for the detection of bacteria than manual microscopic methods [21].

In comparison to the relatively straightforward detection of bacteria in urine samples, blood is a much more challenging sample type to use. Blood has an extremely high cellular concentration (3.2×10^9 to 6.5×10^9 red blood cells.ml^{-1} [22] depending on sex, age etc). In comparison, in clinical conditions such as bacteraemia (the presence of bacteria in blood) concentrations of the contamination may be of the order of 10 bacteria in 1 ml of blood, making detection by microscopic methods all but impossible. Thus, although bacteraemia is a potentially life-threatening condition, diagnosis relies in many cases upon the growth of bacteria in media inoculated with samples of whole blood; this process may take up to 7 days incubation before the results are apparent [23]. However, by selectively lysing the eukaryotic cells in a blood sample, a sufficiently low cell concentration was achieved to allow the rapid sample throughput capabilities of the flow cytometer to be utilised. Using this method Mansour and colleagues were able to detect *Escherichia coli* in blood at concentrations of 10–100 cells.ml^{-1}. This sensitivity was 100 to 1000-fold better than that possible with microscopical methods, and took just 2 hours including sample preparation. In clinical presentations where bacterial concentrations are less than 10 per ml, a short pre-incubation step prior to flow cytometric analysis may be envisaged to increase the bacterial load of the sample to a level where it may be detected.

The detection of specific pathogenic microorganisms in clinical samples has been much improved by the availability of monoclonal antibodies. These antibodies can be labelled directly or indirectly with a range of different fluorochromes to enable them to be detected flow cytometrically. This has proven to be useful for detection of mycobacterial species from clinical (sputum) specimens [24]. Mycobacteria grow very slowly in laboratory culture and therefore a detection method that does not rely on growth is very advantageous for clinical diagnosis purposes. Whilst the antibody used by Yi and colleagues only allowed identification to the genus level, they noted that the ability to distinguish between different species of Mycobacteria was dependent only on the availability of a source of fluorescently-labelled monoclonal antibody. The use of fluorescently-labelled antibodies for the identification of bacteria from clinical isolates to the species level has been demonstrated by Shi and colleagues [25] who used three highly species-specific monoclonal IgG antibodies targeted against *Streptococcus mutans*. *S. mutans* is involved in the initiation and progression of human dental caries and its detection and quantitation are important factors in diagnosis and risk assessment.

6. FCM for detection of microorganisms in the environment

It is widely recognised, especially with samples collected from natural environments, that the total cell count may far exceed the 'viable' count obtained by plating a sample on solidified nutrient media [26, 27]. While some cells may grow slowly, others may fail to grow at all, making their detection and identification by this method impossible. Apart from the obvious explanation of cell death, reasons for lack of colony formation include competition, inhibition, inappropriate media selection etc. FCM permits the detection of cells without any requirement that they can grow. With environmental samples where previously unidentified species may be present (whose cultural requirements are of course unknown) direct rather than culture-based techniques are highly desirable.

Fluorescently-labelled oligonucleotide probes (so called "phylogenetic stains") can be used in conjunction with FCM to provide a high resolution, automated analysis method for the identification of microorganisms. Flow cytometric analysis of microbial samples labelled with probes complementary to group-specific regions of the 16S ribosomal RNA molecule has proved successful [28, 29]. Using FCM to quantify probe-conferred cell fluorescence it was possible to discriminate between target and non-target cell populations of laboratory-grown organisms. However, in nutrient-limited environments where cells may grow slowly or become dormant, the ribosome number may be decreased below the threshold of detection. In these cases signal amplification methods will be helpful [30].

The detection of pathogenic microorganisms in the environment is important for the prevention of disease and screening of water supplies for certain organisms is routine in many countries. *Cryptosporidium parvum* is a microbial parasite that is responsible for outbreaks of the waterborne disease cryptosporidiosis. *Cryptosporidium* oocysts resist conventional water treatment such as chlorination and can persist (and remain infective) for long periods in water. As a consequence of the potentially fatal nature of this disease screening methods for Cryptosporidia are necessary. Flow cytometric methods for the detection of *Cryptosporidium* oocysts have been developed over the last 10 years [31] and now a two-colour, antibody-based detection method is available that reduces the level of false positives [32].

The bacterium *Legionella pneumophila* lives freely in water but can cause Legionnaires' disease when water droplets containing the organism are aerosolised into very fine particles and inhaled deep into the lungs. Cooling towers are often a source of infection, as they produce a very fine aerosol. Whilst the disease can usually be treated effectively with antibiotics, fatality rates can be as high as 10–15%. In the early 1980s a FITC-labelled globulin was shown to bind specifically to Legionella and the resulting fluorescence was detectable by FCM [33]. The applicability of the flow cytometric technique to the detection of *Legionella* spp. in cooling tower water samples was demonstrated by Tyndall and colleagues [34] who used both FITC-labelled antibody plus propidium iodide to stain the bacteria. Using these methods, detection of target bacteria in unconcentrated samples of cooling tower water was possible. The lack of concentration step, together with the rapid nature of flow cytometric analysis resulted in faster and less subjective analysis than traditional microscope-based methods.

The above examples serve to illustrate the role of FCM in screening environmental samples for 'naturally-occurring' contamination. There is currently much scientific and general interest in the detection of deliberate environmental contamination via biowarfare or bioterrorism. A number of bacterial, viral, fungal and toxin agents have been considered as potential threats. However, with its short incubation time (1–3 days) and high mortality rate if left undetected/untreated (95%+ for pulmonary anthrax) *Bacillus anthracis*, the causative organism of anthrax, is the most credible threat [35].

FCM has been used to detect biowarfare agents (or simulants) in the laboratory and during field trials. In the case of anthrax, spores can generally be distinguished from other microorganisms since fluorescent stains do not penetrate the spore coat effectively. In the event of a suspected biowarfare/bioterrorist attack, non-specific stains such as fluorescent brighteners [36] may be sufficient as a cheap first stage screen of environmental samples (Figure 2). However, with time-critical assays a multiparametric flow cytometric protocol combining fluorescently-tagged antibodies with less specific stains could identify target microorganisms in 5 minutes [37]. Further work showed that flow cytometric light scatter data could be used to give an indication of spore viability [38]. Whilst research into the use of FCM for monitoring of the environment

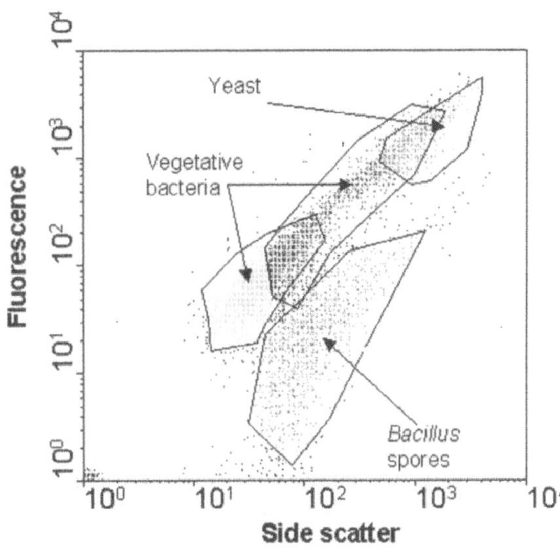

Figure 2. Flow cytometric analysis of a mixture of *B. globigii* spores (simulant for *B. anthracis*), *Escherichia coli*, *Micrococcus luteus* and *Saccharomyces cerevisiae* stained with 40 $\mu g.ml^{-1}$ Tinopal CBS-X. Fluorescence resulted from excitation with a HeCd laser (325 nm) and was collected below 440 nm. Side scatter resulted from excitation with the argon ion laser (488 nm).

can be performed using standard instruments, application of the technique requires specialised, portable instruments. A number of these devices are now becoming available commercially [see e.g. 39] and in some cases can run from their own internal power supply and recycle sheath fluid (via a filter) allowing their operation in remote locations.

7. Concluding remarks

The applications of FCM for the detection of microorganisms are many and varied. The single-cell nature of the analysis method, together with the rapidity of the technique make it ideal for detection of target cells against a background of biological and non-biological particulates. Developments that bring about smaller, cheaper and more robust instrumentation will enable the flow cytometer to become an important tool for the detection of microorganisms for the purposes of safeguarding public health.

References

1. Howlett NG, Avery SV (1999). Flow cytometric investigation of heterogeneous copper-sensitivity in asynchronously grown *Saccharomyces cerevisiae*. FEMS Microbiol Lett 176(2): 379–386.
2. Campbell A, Robertson L, Smith H (1993). Novel methodology for the detection of *Cryptosporidium parvum* – a comparison of cooled charge coupled

devices (CCD) and flow cytometry. Water Science and Technology 27(3–4): 89–92.
3. Nebe-von-Caron G, Stephens P, Badley RA (1998). Assessment of bacterial viability status by flow cytometry and single cell sorting. J Appl Microbiol 84(6): 988–998.
4. Davey HM, Kaprelyants AS, Weichart DH, Kell DB (1999). Approaches to the estimation of microbial viability using flow cytometry. In: Current Protocols in Cytometry. New York: Wiley, pp 11.3.1–11.3.20.
5. Valdivia RH, Falkow S (1996). Bacterial genetics by flow cytometry: rapid isolation of *Salmonella typhimurium* acid-inducible promoters by differential fluorescence induction. Mol Microbiol 22(2): 367–378.
6. Betz JW, Aretz W, Hartel W (1984). Use of flow cytometry in industrial microbiology for strain improvement programs. Cytometry 5: 145–150.
7. Bell PJL, Deere D, Shen J, Chapman B, Bissinger PH, Attfield PV, Veal DA (1998). A flow cytometric method for rapid selection of novel industrial yeast hybrids. Appl Environ Microbiol 64(5): 1669–1672.
8. Davey HM, Kell DB (1996). Flow cytometry and cell sorting of heterogeneous microbial populations – the importance of single-cell analyses. Microbiol Rev 60(4): 641–696.
9. Alvarez-Barrientos A, Arroyo J, Canton R, Nombela C, Sanchez-Perez M (2000). Applications of flow cytometry to clinical microbiology. Clin Microbiol Rev 13(2): 167–195.
10. Vives-Rego J, Lebaron P, Nebe-von Caron G (2000). Current and future applications of flow cytometry in aquatic microbiology. FEMS Microbiol Rev 24(4): 429–448.
11. Harris CM, Kell DB (1985). The estimation of microbial biomass. Biosensors 1: 17–84.
12. Cantinieaux B, Courtoy P, Fondu P (1993). Accurate flow cytometric measurement of bacteria concentrations. Pathobiology 61: 95–97.
13. Veal DA, Deere D, Ferrari B, Piper J, Attfield PV (2000). Fluorescence staining and flow cytometry for monitoring microbial cells. J Immunol Methods 243(1–2): 191–210.
14. Pettipher GL (1991). Preliminary evaluation of flow cytometry for the detection of yeasts in soft drinks. Lett Appl Microbiol 12: 109–112.
15. Gunasekera TS, Attfield PV, Veal DA (2000). A flow cytometry method for rapid detection and enumeration of total bacteria in milk. Appl Environ Microbiol 66(3): 1228–1232.
16. Jimenez L (2001). Rapid methods for the microbiological surveillance of pharmaceuticals. PDA J Pharm Sci Technol 55(5): 278–285.
17. Jespersen L, Lassen S, Jakobsen M (1993). Flow cytometric detection of wild yeast in lager breweries. Int J Food Microbiol 17: 321–328.
18. Attfield PV, Kletsas S, Veal DA, van Rooijen R, Bell PJL (2000). Use of flow cytometry to monitor cell damage and predict fermentation activity of dried yeasts. J Appl Microbiol 89(2): 207–214.
19. Hewitt CJ, Nebe-Von-Caron G (2001). An industrial application of multiparameter flow cytometry: Assessment of cell physiological state and its application to the study of microbial fermentations. Cytometry 44(3): 179–187.
20. Delanghe JR, Kouri TT, Huber AR, Hannemann-Pohl K, Guder WG, Lun A, Sinha P, Stamminger G, Beier L (2000). The role of automated urine particle flow cytometry in clinical practice. Clin Chim Acta 301(1–2): 1–18.
21. Hannemann-Pohl K, Kampf SC (1999). Automation of urine sediment examination: A comparison of the sysmex UF-100 automated flow cytometer with routine manual diagnosis (microscopy, test strips, and bacterial culture). Clinical Chemistry and Laboratory Medicine 37(7): 753–764.
22. Eastham RD (1984). Clinical haematology, 6th ed. Bristol: John Wright and Sons.
23. Mansour JD, Robson JA, Arndt CW, Schulte TE (1985). Detection of *Escherichia coli* in blood using flow cytometry. Cytometry 6: 186–190.
24. Yi WC, Hsiao S, Liu JH, Soo PC, Horng YT, Tsai WC, Lai HC, Teng LJ, Hsueh PR, Hsieh RF, Luh KT, Ho SW (1998). Use of fluorescein labelled antibody and fluorescence activated cell sorter for rapid identification of Mycobacterium species. Biochem Biophys Res Commun 250(2): 403–408.
25. Shi W, Jewett A, Hume WR (1998). Rapid and quantitative detection of *Streptococcus mutans* with species-specific monoclonal antibodies. Hybridoma 17(4): 365–371.
26. Hugenholtz P, Pace NR (1996). Identifying microbial diversity in the natural environment: a molecular phylogenetic approach. Trends Biotech 14(6): 190–197.
27. Amann RI, Ludwig W, Schleifer KH (1995). Phylogenetic identification and *in situ* detection of individual microbial cells without cultivation. Microbiol Rev 59(1): 143–169.
28. Wallner H, Amann R, Beisker W (1993). Optimizing fluorescent *in situ* hubridization with rRNA-targeted oligonucleotide probes for flow cytometric identification of microorganisms. Cytometry 14: 136–143.
29. Amann RI, Binder BJ, Olson RJ, Chisholm SW, Devereux R, Stahl DA (1990). Combination of 16S rRNA-targeted oligonucleotide probes with flow cytometry for analyzing mixed microbial populations. Appl Environ Microbiol 56(6): 1919–1925.
30. Zarda B, Amann R, Wallner G, Schleifer KH (1991). Identification of single bacterial-cells using digoxigenin-labeled, ribosomal-RNA-targeted oligonucleotides. J Gen Microbiol 137(Pt12): 2823–2830.
31. Vesey G, Slade JS, Byrne M, Shepherd K, Fricker CR (1993). A new method for the concentration of *Cryptosporidium* oocysts from water. J Appl Bacteriol 75: 82–86.
32. Ferrari BC, Vesey G, Davis KA, Gauci M, Veal D (2000). A novel two-color flow cytometric assay for the detection of Cryptosporidium in environmental water samples. Cytometry 41(3): 216–222.
33. Ingram M, Cleary TJ, Price BJ, Price RL, Castro A (1982). Rapid detection of *Legionella pneumophila* by flow cytometry. Cytometry 3(2): 134–147.
34. Tyndall RL, Hand Jr. RE, Mann RC, Evans C, Jeringen R (1985). Application of flow cytometry to detection and characterization of *Legionella* spp. Appl Environ Microbiol 49(4): 852–857.

35. Dando M (1994). Biological warfare in the 21st century. London: Brassey's.
36. Davey HM, Kell DB (1997). Fluorescent brighteners: Novel stains for the flow cytometric analysis of microorganisms. Cytometry 28(4): 311–315.
37. Sincock SA, Kulaga H, Cain M, Anderson P, Stopa PJ (1999). Applications of flow cytometry for the detection and characterization of biological aerosols. Field Analytical Chemistry and Technology 3(4–5): 291–306.
38. Stopa PJ (2000). The flow cytometry of *Bacillus anthracis* spores revisited. Cytometry 41(4): 237–244.
39. Davey HM, Kell DB (2000). A portable flow cytometer for the detection and identification of microorganisms. In: Stopa PJ, Bartoszcze MA (eds), Rapid Methods for Monitoring the Environment for Biological Hazards. Dordrecht: Kluwer, pp 159–167.

Address for correspondence: Hazel M. Davey, Institute of Biological Sciences, University of Wales, Aberystwyth, Ceredigion, SY23 3DD Wales, UK
Phone: +44 (0) 1970 621829; Fax: +44 (0) 1970 622307
E-mail: hlr@aber.ac.uk;
Web http://qbab.aber.ac.uk/home.html

Methods in Cell Science **24**: 99–106 (2002).

A flow cytometric approach to assess phytoplankton respiration

Gérald Grégori[1], Michel Denis[1], Dominique Lefèvre[1] & Beatriz Beker[2]

[1] *Laboratoire d'Océanographie et de Biogéochimie, Université de la Méditerranée, CNRS UMR 6535, 163 avenue de Luminy, Case 901, 13288 Marseille cedex 9, France;* [2] *Centre d'Océanologie de Marseille, CNRS UMS 2196, Station Marine d'Endoume, rue de la Batterie des Lions, 13007 Marseille, France*

Abstract. Microbial respiration in the ocean is considered as the major process representative of the organic matter biological oxidation. The corresponding metabolic CO_2 production was estimated to be about 22 Pg C y^{-1} [24]. However, the *in situ* respiration rate is generally too low (by several orders of magnitude) to be accessible to the available direct measurement methods. Some fluorescent probes, such as $DiOC_6(3)$ (Molecular Probes, USA) have been shown to be very sensitive to changes in the proton electrochemical potential difference ($\Delta\tilde{\mu}_{H+}$), characterising mitochondrial and plasmic membranes bearing the cell respiratory system in eukaryotic and prokaryotic cells respectively [22, 37]. In mitochondria, $\Delta\tilde{\mu}_{H+}$ is linked to the flux of oxygen uptake by a linear relationship [32]. To our knowledge, no such relationship has been established in the case of whole marine cells. In the present work, we addressed the dark respiration rate of the Chlorophyceae *Dunaliella tertiolecta* (Butcher) in axenic cultures, both directly by using a highly sensitive oxygraph (Oroboros) and by staining cells with $DiOC_6(3)$. We found and standardized a linear relationship between oxygen uptake by *D. tertiolecta* and its green fluorescence induced by $DiOC_6(3)$, enabling the determination by flow cytometry of the respiration rate of *D. tertiolecta*.

Key words: Aquatic microorganisms, Cell respiration, $DiOC_6(3)$, Flow cytometry

1. Introduction

Cellular respiration is the ultimate step in the organic matter biological oxidation. Under aerobic conditions, the most general case in the ocean water column, cell respiration results in consumption of molecular dioxygene (O_2), production of carbon dioxide (CO_2) and energy in the form of adenosine triphosphate (ATP). The carbon, initially removed from the natural environment by autotrophic organisms *via* photosynthesis, is thus released into the Ocean, and could further contribute to the green house effect after diffusion to the atmosphere. The oceanic budget for this metabolic CO_2 production was estimated to be around 22 Pg C y^{-1} [24]. If the metabolic CO_2 production is difficult to quantify by direct measurements, it can be assessed more easily by measuring the cellular respiration rate.

However, due to the very low activity levels occurring *in situ*, the currently available direct techniques are not appropriate to their determination. Consequently, indirect methods have been developed. Chemical techniques address bulk changes in dissolved oxygen concentrations (Winkler method; [1]) or in dissolved inorganic carbon (coulometry; [13]). These methods usually require long-term incubations (typically 24 hours) for sea-surface samples and are not sensitive enough to be applied to samples from the aphotic layer due to lower cell concentrations and metabolisms. In the case of O_2 consumption measurements, data have to be expressed in terms of metabolic CO_2 production by using conversion coefficients [31, 39].

The enzymatic approach developed by Packard [23] quantifies the maximum oxygen uptake rate linked to the electron flux through the dehydrogenases embedded in the bioenergetic membranes. This very sensitive method also requires conversion factors to determine the *in situ* oxygen uptake rate. Another enzymatic approach has been developed to directly address the decarboxylation activity of the isocitrate-dehydrogenase in the Krebs cycle [25, 26, 35]. The main interest in quantifying directly the CO_2 metabolic production is to avoid the O_2/CO_2 conversion step which is a source of additional uncertainties. However, this enzymatic approach also delivers potential activities that need to be converted to physiological rates. Up to now, there has been no application to natural samples.

To face this lack of direct measurement method for *in situ* microbial respiratory activity, we investigated the capacity of flow cytometry combined with the use of fluorescent probes sensitive to membrane potential. Studies on mitochondrial bioenergetics have shown that there is a linear relationship [32] between the oxygen flux and the proton electrochemical potential difference ($\Delta\tilde{\mu}_{H+}$) with respect to both sides of the inner mitochondrial membrane. The coupling between the electron transfer through the respiratory system up to molecular oxygen reduction to water, and the proton translocation by which the energy conversion occurs is accounted for by the chemiosmotic theory of Mitchell (1961; see also Nicholls, 1982 [21]). Several $\Delta\tilde{\mu}_{H+}$ sensitive molecular dyes, such as Rhodamine 123, have been successfully experimented with purified mitochondria [17, 27, 28, 33] or isolated cells [5, 15, 29, 33] but few published works concern living cells from the marine environment [3, 4]. To our knowledge, no relationship between oxygen flux and $\Delta\tilde{\mu}_{H+}$ has been established in the case of whole cells. It was therefore highly relevant (i) to stain a cell suspension with a fluorescent molecular probe sensitve to $\Delta\tilde{\mu}_{H+}$, (ii) to quantify by flow cytometry the fluorescence intensity of the stained cells (representative of $\Delta\tilde{\mu}_{H+}$) and (iii) to independently measure the O_2 uptake rate by the cell suspension in order to establish the relationship between both measurements. The experiments were conducted on axenic cultures of *Dunaliella tertiolecta* (Chlorophyceae) and we used $DiOC_6(3)$ as the fluorescent dye. Results provided evidence that the relationship for *D. tertiolecta* is linear, enabling in a simple way to convert fluorescence signals into respiration rates. This conversion has been standardized so that derived respiration rates are independent from the flow cytometer being used.

2. Materials and procedures

Phytoplankton culture

The unicellular organism chosen for the present work was the Chlorophyceae *Dunaliella tertiolecta* (Butcher). This ovoid (6–8 µm in diameter) marine phytoplankton, is easy to grow and very suitable for laboratory studies. Two locomotive flagella maintain cells in suspension, preventing any cell loss on the container wall and additional errors in cell counts by flow cytometry. The *D. tertiolecta* strain (CS-175) used in this study was from the CSIRO Marine Research Culture Collection (Hobart, Australia). It was grown under axenic conditions on f/2 *medium* [10] at 20 °C, a salinity of 33 and with an irradiance of 100 $\mu E\ m^{-2}\ s^{-1}$ provided by fluorescent OSRAM lamps with a 12 h light-dark cycle.

All experiments were run with cells in the exponential growth phase and kept in the dark in order to address dark respiration and to prevent O_2 production that would occur if photosynthesis was activated.

Fluorescent probe

The fluorescent dye used to monitor the cellular respiration by flow cytometry was the carbocyanine dihexyloxacarbocyanine iodide ($DiOC_6(3)$, Molecular Probes, USA; [11]). This hydrophobic molecule is intrinsically fluorescent (*maximum* excitation at 484 nm; *maximum* emission at 501 nm) but its quantum yield is typically 4 to 6 times higher in non-polar environment than in aqueous solution [38]. Working $DiOC_6(3)$ solutions (100 µM) were prepared in DMSO (Sigma Chemical Co., USA) and could be kept several weeks in the dark, at room temperature.

Uncoupler

The carbonyl cyanide m-chloro phenyl hydrazone (CCCP, Sigma, USA) was used to titrate the uncoupling of mitochondrial electron transfer from proton translocation. The CCCP working solution (3 mM) was prepared in ethanol. Variations in uncoupling efficiency were observed between different CCCP batches in terms of reaction time and extent of the uncoupling. In order to compensate for the reaction-time variations, a 10 min incubation time was applied to each CCCP addition.

High-resolution respirometry

Respiration rates were recorded with the high-resolution oxygraph OROBOROS (Paar, Innsbruck, Austria), equipped with two measurement chambers [6–8]. The culture medium in the open chamber kept at 20 °C (temperature controlled by Pelletier effect) and homogenized with a magnetic stirrer (600 rpm), was left to equilibrate with the atmosphere before signal calibration. Oxygen concentration was computed from the partial pressure measured by the electrode. During respiration measurements, each chamber (3 cm^3 working volume) was closed by a piston to prevent oxygen exchange with the atmosphere. Data were recorded on a desktop computer every 0.4 s by using the Datlab Acquisition software (Paar, Innsbruck, Austria). Data analysis was run with the Datlab Analysis software (Paar, Innsbruck, Austria) to calculate oxygen consumption rates ($\mu M.h^{-1}$), and correct them from the electrode shift. The OROBOROS precision for O_2 concentration measurements is 0.036 µM against 0.500 to 1.000 µM for other commercial instruments.

Flow cytometry

A CYTORON ABSOLUTE (ORTHO Diagnostic Systems) flow cytometer, equipped with an air-cooled argon laser (excitation wavelength at 488 nm) was used for single cell analysis. Each cell was characterized by 5 optical parameters: two scatter parameters, namely forward angle scatter (related to the particle size) and right angle scatter (related to cell structure and shape), and three fluorescence parameters related to emissions in the red (> 620 nm), orange (565–592 nm) and green (515–530 nm) wavelength ranges. Data were collected and stored in list-mode on a personal computer with the IMMUNOCOUNT software (ORTHO Diagnostic

Systems). Due to the control of the sample volume by a microsyringe, this software can directly provide cell concentrations (cells mm^{-3}). The Winlist software (Verity Software, USA) was used to determine statistical data.

3. Results

Titration of D. tertiolecta respiration

The search for a relationship between the rate of respiration and the cell fluorescence induced by a probe sensitive to $\Delta\tilde{\mu}_{H+}$ implies the capacity to monitor the *D. tertiolecta* respiration rate over a range representative of its natural variations. To achieve this goal, titration of the respiration rate of *D. tertiolecta* was processed by addition of increasing amounts of CCCP, an uncoupler whose effect results in stimulating electron transfer and consequently, dioxygen uptake [2, 12, 34].

Control experiments presented in Figure 1 showed that CCCP addition stimulated mitochondrial respiration rate in a way stable and long enough to be measured with the oxygraph and to enable the recording of the fluorescence level of *D. tertiolecta* cells stained with $DiOC_6(3)$. Under the conditions of Figure 1 where mitochondria were not fully uncoupled, the respiration rate was increased 2.4 fold. The observed maximum enhancement corresponding to fully uncoupled mitochondria was 3.8 fold (case of Figure 6).

Increasing values of O_2 uptake rate by *D. tertiolecta* were determined with the oxygraph after addition of increasing amounts of CCCP. Results displayed in Figure 2 reveal a linear relationship between CCCP concentration and O_2 consumption rate by *D. tertiolecta*.

Staining of D. tertiolecta

As described in the Materials and Methods section, the fluorescent molecular probe $DiOC_6(3)$ was used to monitor $\Delta\tilde{\mu}_{H+}$ variations. The potential impact of $DiOC_6(3)$ staining on *D. tertiolecta* metabolism was checked by measuring cell respiration when supplementing culture samples with dye concentrations in the range 100–1,000 nM. This range of concentrations was explored by successive additions of 3 mm^3 of the 100 µM $DiOC_6(3)$ working solution (Figure 3A). The O_2 consumption rates determined after each $DiOC_6(3)$ addition are reported in Figure 3B. The resulting respiratory rates (47.1 ± 5.8 µmol O_2 dm^{-3} h^{-1}) were found not significantly different (10 measurements; $p < 0.05$) from that (46.1 ± 2.1 µmol O_2 dm^{-3} h^{-1}) in absence of dye over the extent of the experiment (Figure 3A).

The optimal dye concentration is expected to yield a strong and stable fluorescence signal providing a sensitive response to changes in mitochondrial $\Delta\tilde{\mu}_{H+}$. In order to prevent any contribution from chloroplast bioenergetics, samples were kept in darkness. Kinetics of *D. tertiolecta* staining with 10 to 90 nM $DiOC_6(3)$ (final concentrations) were recorded over 1 h periods. The corresponding time courses of green fluorescence intensities are displayed in Figure 4.

A fast phase staining, expressed by a fast rise in green fluorescence, occurred within the first 5 minutes following the dye addition, whatever the concentration used. For $DiOC_6(3)$ concentrations = 30 nM, the green fluorescence intensity remained stable after 5 min incubation. In contrast, above 30 nM $DiOC_6(3)$, the green fluorescence signal increased continuously. Consequently, $DiOC_6(3)$ was used at 30 nM concentration to monitor $\Delta\tilde{\mu}_{H+}$ when titrating *D. tertiolecta* respiration rate. In order to ensure optimum staining, *D. tertiolecta* cells were left 30 min in presence of $DiOC_6(3)$ before flow cyto-

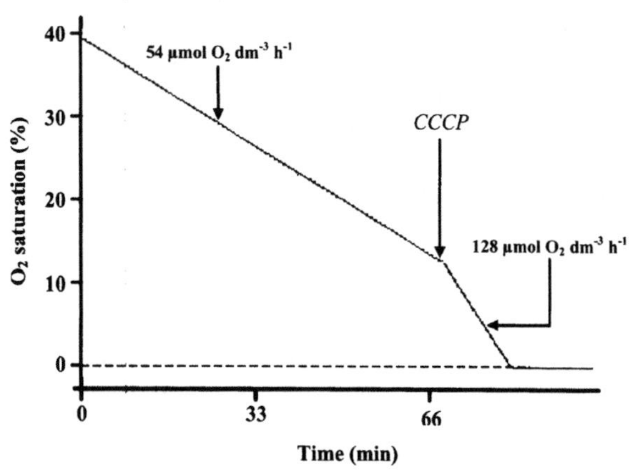

Figure 1. Action of CCCP on molecular oxygen uptake rate in a sample of *D. tertiolecta* culture. Addition of 5 mm^3 of CCCP (3 mM) stimulated the O_2 uptake rate by about 2.4 fold.

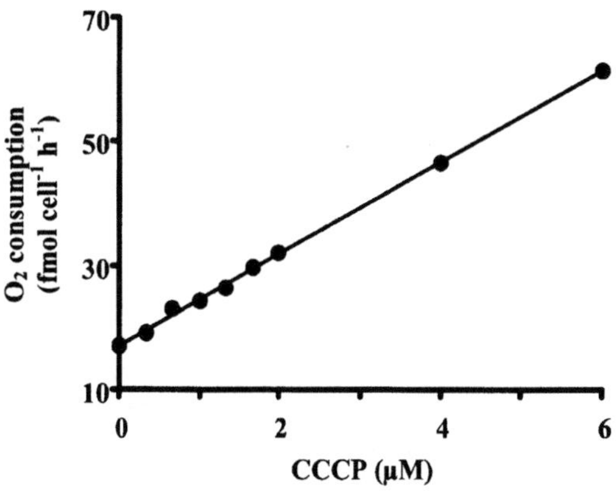

Figure 2. Relationship between the rate of O_2 consumption by *D. tertiolecta* and CCCP concentration. The straight line expresses a linear relationship between O_2 uptake rate and CCCP concentration: $y = 2.45x + 17.18$; $R^2 = 0.99$; n = 9.

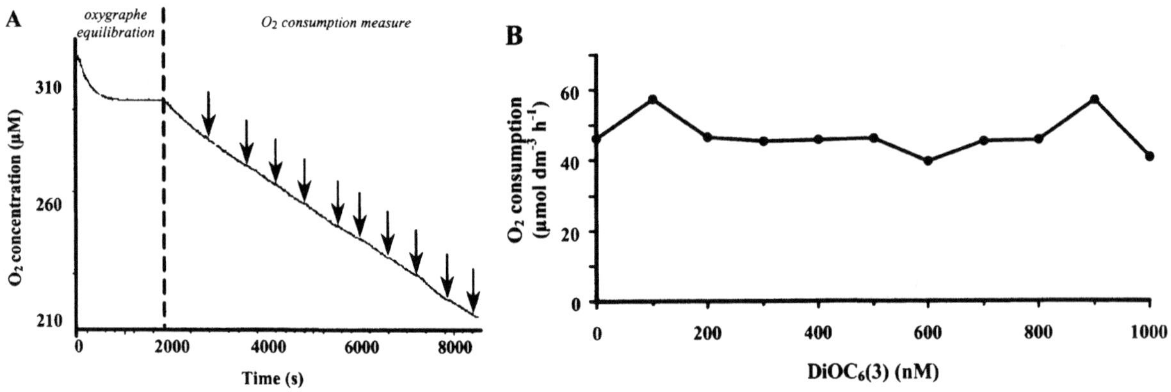

Figure 3. **(A)** Effect of $DiOC_6(3)$ staining of *D. tertiolecta* in the course of respiration measurements. Each arrow represents the addition of 3 mm^3 of the $DiOC_6(3)$ working solution (100 μM), increasing the dye concentration by 100 nM steps, up to 1 μM. **(B)** Variations of *D. tertiolecta* O_2 uptake rate upon increasing $DiOC_6(3)$ staining.

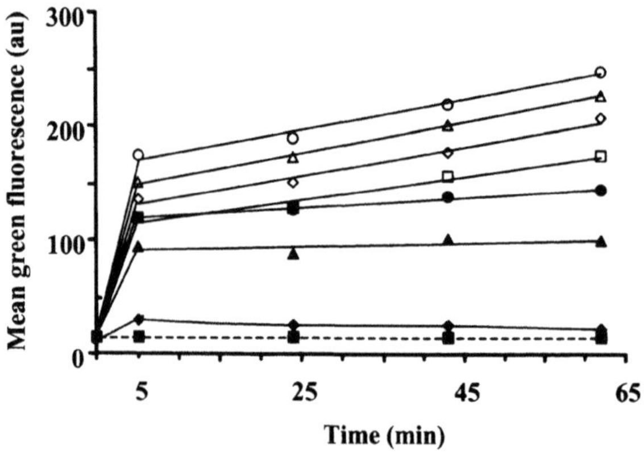

Figure 4. Green fluorescence kinetics of *D. tertiolecta* cells in absence of dye (■) and for different $DiOC_6(3)$ additions: 10 (◆), 30 (▲), 50 (●), 60 (□), 70 (◇), 80 (△), 90 (○) nM. The time scale refers to the time elapsed after dye addition.

metric analysis, in agreement with similar incubation times reported in the literature [19, 34].

D. tertiolecta staining was controlled by epifluorescence microscopy (not shown) to localize the fluorochrome binding sites within the cells and check for possible extramitochondrial distribution of the dye that can be seen as a diffused green fluorescence for $DiOC_6(3)$ concentrations > 50 nM.

Monitoring of respiration

Subsamples of *D. tertiolecta* pretreated with increasing amounts of CCP like in the above experiment, were incubated 30 min with $DiOC_6(3)$ at 30 nM final concentration before flow cytometric analysis. As expected, the decrease in $\Delta\tilde{\mu}_{H+}$ induced by the added CCCP resulted in the decrease of the green fluorescence signal of the stained cells as shown in Figure 5A. This monitoring of $\Delta\tilde{\mu}_{H+}$ yielded a linear relationship between CCCP concentration and the cell mean green-fluorescence intensity as reported in Figure 5B. When mitochondria were fully uncoupled, the cell mean-green fluorescence intensity was similar to that of cells without staining (about 4 a.u., Figure 5B).

Conversion scale

The two linear relationships obtained independently between CCCP concentration and *D. tertiolecta* respiration rate on the one hand and, CCCP concentration and green fluorescence of stained *D. tertiolecta* on the other hand, demonstrate the existence of a linear relationship between *D. tertiolecta* respiration rate and its green fluorescence when stained with $DiOC_6(3)$. This linear relationship can be easily constructed from two specific values of *D. tertiolecta* respiration rate and the associated green fluorescence of cells in the exponential growth phase, stained with $DiOC_6(3)$ when in fully coupled and fully uncoupled state respectively. Both respiration rates can be considered as specific features of *D. tertiolecta* in its exponential growth phase: namely 17 and 64 fmol O_2 $cell^{-1}$ h^{-1} for cells in fully coupled and fully uncoupled state respectively. However, the associated fluorescence values depend on the instrument being used. The linear scale conversion corresponding to our instrument is plotted in Figure 6. The green fluorescence values were corrected from the residual green fluorescence recorded in the absence of $DiOC_6(3)$.

Application

To check the efficiency of the linear conversion scale, an experiment was run in which the respiration rate of *D. tertiolecta* was measured directly with subsamples supplemented with CCCP (10 min) then with $DiOC_6(3)$ (30 min). At the end of the incubation time, the cell suspension was removed from the oxygraph chamber to be analyzed by flow cytometry. The cell mean green fluorescence for each respiratory stage

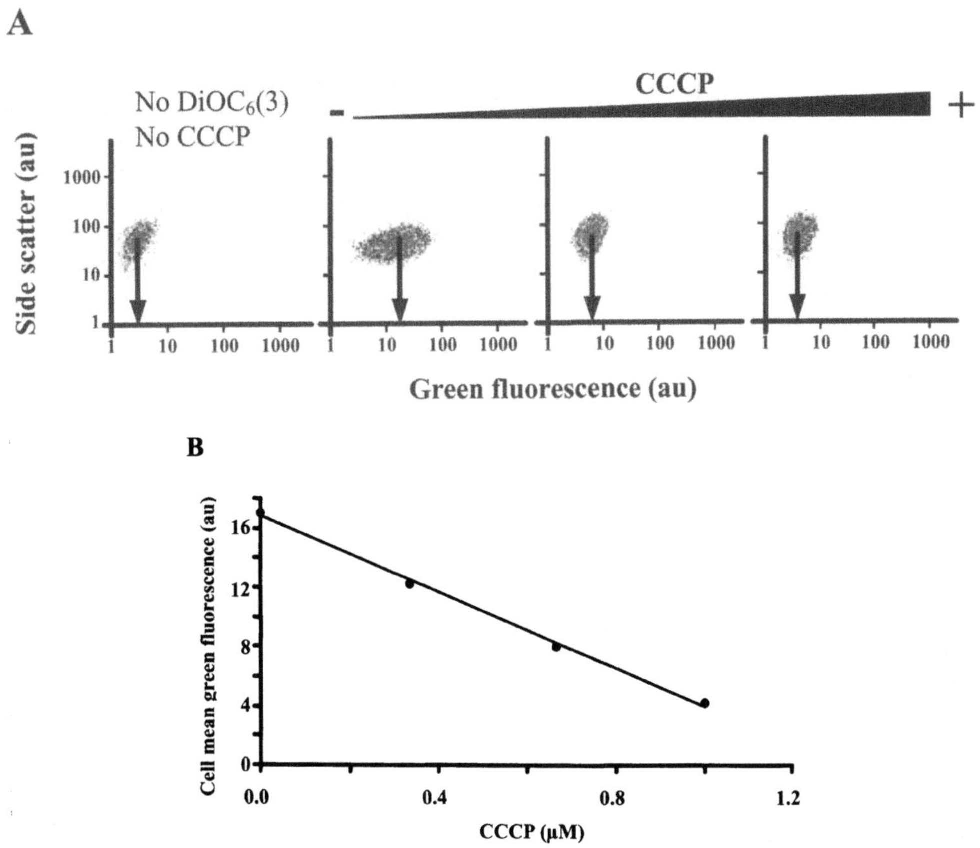

Figure 5. **(A)** Effect of CCCP on *D. tertiolecta* staining with $DiOC_6(3)$. Increasing amount of CCCP (black arrows) reduced the green fluorescence level of the cells, as expected from a decreasing $\Delta\tilde{\mu}_{H+}$. Note that the dye was supplemented after CCCP addition. **(B)** Relationship between *D. tertiolecta* cell mean green fluorescence intensity and CCCP concentration. The straight line expresses a linear relationship between the cell mean green fluorescence intensity (in arbitrary units, au) and CCCP concentration: $y = -12.82x + 16.82$; $R^2 = 0.997$; n = 4.

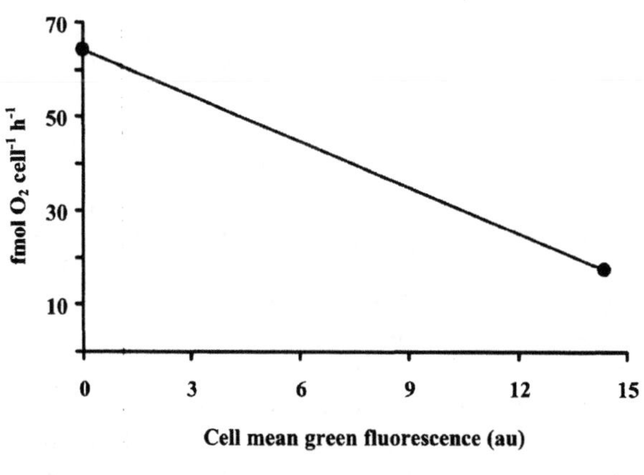

Figure 6. Linear conversion scale between the cell O_2 respiration rate and the cell mean green fluorescence intensity (in arbitrary units, au) for *D. tertiolecta* stained with 30 nM $DiOC_6(3)$. The linear scale is based on two specific respiration rates of *D. tertiolecta* in exponential growth phase, corresponding to fully coupled and fully uncoupled states respectively, and the associated fluorescence intensities. The resulting linear equation, $y = -3.25x + 64.09$, is therefore only valid for the instrument in use. To make this relationship instrument-independent would require a standardization of fluorescence measurements.

was corrected from the residual green fluorescence determined in the absence of $DiOC_6(3)$ and then converted to respiration rate by means of the linear relationship shown in Figure 6. Calculated and measured respiration rates appeared not significantly different within the experimental uncertainties as shown in Figure 7 ($R^2 = 0.984$; 21 measures).

4. Discussion

Microorganism cell respiration is considered as representative of organic matter remineralisation in the ocean. The biological oxidation activity has been estimated to amount to 22 Pg C y^{-1} [24], demonstrating the importance of this process at the ocean scale. However, the *in situ* respiration rate is too low to be directly measured by currently available techniques. The indirect chemical and enzymatic methods that have been developed require large incubation times and/or conversion factors that fuel unending controversial debates. Over the past decades, the research emphasis was concentrated on the production of organic matter [41]. Far less efforts were

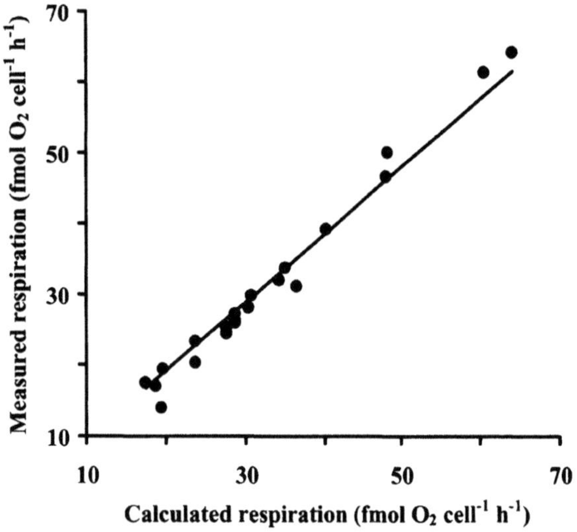

Figure 7. Comparison between measured and calculated mean cell respirations. The reported *D. tertiolecta* cell mean respiration rates were measured directly with the oxygraph and derived from the green fluorescence recorded by flow cytometry by using the linear scale conversion plotted in Figure 6. The straight line demonstrates a tight correlation between both data sets ($y = 0.96x$, $R^2 = 0.98$, n = 21).

dedicated to its fate. It is therefore timely to investigate more intensively this complementary domain of the carbon cycle to achieve a better quantification and understanding of the output of the biological pump with respect to carbon. In the present study, we propose a direct approach based on the capacity of some fluorescent probes to express the proton electrochemical potential difference ($\Delta\tilde{\mu}_{H+}$) between both sides of bioenergetic membranes (Mitchell, 1961 [21]). To establish the feasibility of this approach, we selected a eukaryotic phytoplanktonic cell, *Dunaliella tertiolecta* considering that (i) this unicellular microalgae does not form aggregates and therefore is very suitable for flow cytometric analyses, (ii) the mobility of this microorganism generated by its two locomotive flagella minimizes sedimentation and sticking on the glass wall and (iii) its natural fluorescence occurs essentially in the red wavelength range and does not overlap with the green fluorescence emission of the selected fluorescent probe, $DiOC_6(3)$, belonging to the carbocyanine family. In addition, this ubiquitous organism is easy to grow in the laboratory.

CCCP

In order to investigate the respiration rate of *D. tertiolecta* over a range of physiological significance, we used the uncoupler CCCP whose effect is to stimulate the cell respiration rate. This small amphiphilic and hydrophobic molecule is able to penetrate phospholipidic membranes and to increase their permeability to protons (H^+) [2, 12, 34], thus bypassing the return of protons to the mitochondrial matrix through the ATPase. The respiration rate enhancement produced by the uncoupling between electron transfer and proton translocation reached a maximum of 2.4 in the experiment displayed in Figure 1 which provided a range wide enough to establish a relevant linear relationship between *D. tertiolecta* respiration rate and CCCP concentration (Figure 2).

$DiOC_6(3)$

$DiOC_6(3)$ is a lipophilic compound, positively charged, that therefore specifically binds bioenergetic inner membranes of mitochondria, the most electronegative (around –100 mV) organelle in the cell. However, when the dye concentration is large (> 1 μM), it accumulates into other lipidic domains like the endoplasmic reticulum [9, 36] or microtubules [16].

Moreover, at high concentration (> 1 μM), this dye can be toxic, inhibiting the NADH-dehydrogenase of the respiratory chain [34]. It is therefore necessary to use $DiOC_6(3)$ with concentrations < 1 μM not only to restrict its binding to the inner mitochondrial membranes but also to prevent any toxic effect on the respiratory chain [34, 38] which is supported by the experiment displayed in Figure 3 where $DiOC_6(3)$ concentrations were in the range 100–1,000 nM.. As shown in Figure 4, *D. tertiolecta* stained with very low $DiOC_6(3)$ concentrations (< 50 nM) can produce, when excited with the 488 nm laser beam, a significant green fluorescence signal that remains stable with time. This is favored by the quantum yield of the dye that increases by a factor 6 when it is in an hydrophobic environment [38]. With $DiOC_6(3)$ concentrations over 50 nM, the green fluorescence intensity continuously increased over the experiment time scale. These different behaviors can be explained by assigning the $DiOC_6(3)$ binding sites almost exclusively to the inner mitochondrial membranes for $DiOC_6(3)$ amounts < 50 nM and to all cell membranes for $DiOC_6(3)$ concentrations > 50 nM. The low rate of fluorescence increase is accounted for by the lower affinity of the dye for non energetic membranes. This interpretation is supported by the action of the uncoupler. Indeed, when cells are fully uncoupled, the green fluorescence observed after staining with 30 nM $DiOC_6(3)$ is identical to that recorded before staining (Figure 5). It is worth noting that the order of chemical additions is critical. Indeed, binding of $DiOC_6(3)$ is irreversible which requires to first add CCCP to stimulate respiration then supplement the cell suspension with $DiOC_6(3)$ to monitor $\Delta\tilde{u}_{H+}$. Proceeding otherwise would yield a green fluorescence signal unrelated to $\Delta\tilde{\mu}_{H+}$. The $DiOC_6(3)$ concentration (30 nM) and incubation time (30 min) chosen for *D. tertiolecta* cultures in concentration of about 10^6 cells cm^{-3} are similar to those available in the literature related to eukaryotic cells (50 nM; 30 min incubation [37].

Monitoring respiration

The low (30 nM) $DiOC_6(3)$ concentration used to monitor $\Delta\tilde{\mu}_{H+}$ makes unnecessary a washing step [14, 30]. The observed linear relationship (Figure 5B) between the green fluorescence of stained cells and CCCP concentration combined with the linear relationship (Figure 2) between *D. tertiolecta* respiration rate and CCCP concentration demonstrates the existence of a linear relationship between *D. tertiolecta* respiration rate and its green fluorescence when stained with 30 nM $DiOC_6(3)$. This linear relationship extends to the whole cell that observed for isolated purified mitochondria [32]. It enables a simple and direct conversion of fluorescence signals into respiration rates. The scale conversion plotted in Figure 6 was built by using two specific respiration rates of *D. tertiolecta* corresponding to fully coupled and fully uncoupled cells in the exponential growth phase respectively. The specificity of these respiration rates is supported by an earlier study of the temperature dependence of *D. tertiolecta* respiration rate for cells in the exponential growth phase. Indeed, in the temperature range 10–35 °C, the respiration rate varied from 10 to 70 fmol O_2 $cell^{-1}$ h^{-1} respectively (unpublished results). Due to the instrument dependence of the fluorescence measurements, a conversion scale like that in Figure 6 ($y = -3.25x + 64.09$) has to be defined for each flow cytometer, the intercalibration between instruments being ensured by the two reference respiration rates.

The independent measurements of respiration rates and fluorescence intensities on the same set of subsamples stimulated with CCCP and stained with $DiOC_6(3)$ lead to calculated respiration rates highly correlated to the measured ones (Figure 7; $y = 0.96x$, $R^2 = 0.984$, n = 21).

The demonstration of the feasibility to derive the respiration rate of *D. tertiolecta* from flow cytometric measurements, open the way to the extension of this approach to the other microorganisms, their bioenergetics, on which this approach is built, being basically the same. However, if there is no theoretical argument to preclude the extension of the present approach, experimental obstacles may obstruct the way ahead. For instance, the stain must reach the bioenergetic membrane and some species containing glycoproteins are known to actively pump out some molecular probes [40] like rhodamine 123 [20], BCECF (from BCECF-AM), carboxyfluorescein (from CFDA) [14]. If staining of mammalian cells is well documented and controlled, the situation is different for algae and marine bacteria [40]. Compared to GRAM positive bacteria, GRAM negative bacteria, for instance, have an additional cell wall which acts as a barrier towards some lipophilic dyes (some cyanines and rhodamine 123 [18, 38, 40]). This obstacle was solved by pretreatments with Tris and EDTA in order to permeabilise the external membrane [38]. Whatever the difficulties ahead, the present study opens the way to a new, elegant and efficient way to quantify cell respiration, a major challenge in oceanography.

Acknowledgements

The authors wish to thank S. Brown and D. Marie for their encouragements and precious advices at the early stage of this work. G. G. is the recepient of a doctoral fellowship from the Council of the Provence-Alpes-Côte-d'Azur Region.

References

1. Aminot A. Dosage de l'oxygène dissous. In: CNEXO, (ed). Manuel des analyses chimiques en milieu marin, pp 75–92. Brest; 1983.
2. Bunting JR, Phan TV, Kamali E, Dowben RM. Fluorescent cationic probes of mitochondria: metrics and mechanism of interaction. J Biophys Soc 1989; 56: 979-993.
3. Cid A, Fidalgo P, Herrero C, Abalde J. Toxic action of copper on the membrane system of a marine diatom measured by flow cytometry. Cytometry 1996; 25: 32–36.
4. Cid A, Herrero C, Abalde J. Functional analysis of phytoplankton by flow cytometry a study of the effect of copper on a marine diatom. Sci Mar 1996; 60 (Supl 1): 303–308.
5. Ehrenberg B, Karkas DL, Fluhler EN, Lojewska Z, Loew LM. Membrane potential induced by external electric field pulses can be followed with a potentiometric dye. Biophys J 1987; 51: 833–837.
6. Gnaiger E, Forstner H. Polarographic sensors, Aquatic and physiological applications. Gnaiger E, Forstner H (eds). Berlin/Heidelberg/New York: Springer-verlag, 1983.
7. Gnaiger E, Kuznetsov AV, Lasnig B, Fuchs A, Reck M, Renner K, et al. High-resolution respirometry – optimum permeabilization of the cell membrane by digitonin. In: Larsson C, Pahlman TL, Gistatsson L, (eds), Bio thermokinetics in the post genomic era, pp 89–95. Göteborg: Chalmers Reproservice, 1998.
8. Gnaiger E, Steinlechner-Maran R, Méndez G, Eberl T, Margreiter R. Control of mitochondrial and cellular respiration by oxygen. J Bioenergetics and Biomembranes 1995; 27(6): 583–596.
9. Gordienko DV, Greenwood IA, Bolton TB. Direct visualization of sarcoplasmic reticulum regions discharging Ca^{2+} sparks in vascular myocytes. Cell Calcium 2001; 29(1): 13–28.
10. Guillard RRL, Ryther JH. Studies of marine planktonic diatoms. Part 1: *Cyclotella nana* Hustedt and *Detonula confervacea* (Cleve) Gran. Can J Microbiol 1962; 8: 229–239.

11. Spence MTZ, ed. Handbook of fluorescent probes and research chemicals. Eugene, Oreg: Molecular Probes, 1998.
12. Heytler PG. Uncouplers of oxidative phosphorylation. In: Erecinska M, Wilson DF (eds), Inhibitors of mitochondrial function: Pergamon Press; 1981: 199–210.
13. Johnson KM, Sieburth JM. Coulometric total carbon dioxide analysis for marine studies: automation and calibration. Mar. Chem. 1987; 21: 117–133.
14. Joux F, Lebaron P. Use of fluorescent probes to assess physiological functions of bacteria at single-cell level. Microbes and Infection 2000; 2: 1523–1535.
15. Juan G, Cavazzoni M, Saez GT, O'Connor JE. A fast kinetic method for assessing mitochondrial membrane potential in isolated hepatocytes with rhodamine 123 and flow cytometry. Cytometry 1994; 15: 335–342.
16. Lee C, Wu SS, Chen LB. Photosensitization by 3,3'-dihexyloxacarbocyanine iodide Specific disruption of microtubules and inactivation of organelle motility. Can Res 1995; 55: 2063–2069.
17. Leprat P, Ratinaud MH, Maftah A, Petit JM, Julien R. Use of nonyl acridine orange and rhodamine 123 to follow biosynthesis and functionnal assembly of mitochondrial membrane during L1210 cell cycle. Exp Cell Res 1990; 186: 130–137.
18. Mc Feters GA, Yu FP, Pyle BH, Stewart PS. Physiological assessment of bacteria using fluorochromes. J Microbiol Methods 1995; 21: 1–13.
19. Monfort P, Ratinaud MH, Got P, Baleux B. Apports de la cytométrie en flux et en images en nécologie bactérienne des milieux aquatiques. Océanis 1995; 21(1): 97–111.
20. Nebe-Von Caron G, Badley RA. Viability assessment of bacteria in mixed populations using flow cytometry. J. Microscopy 1995; 179(1): 55–66.
21. Nicholls DG. Bioenergetics. An introduction to the chemiosmotic theory. London, New York: Academic Press Inc.; 1982.
22. Novo D, Perlmutter NG, Hunt RH, Shapiro HM. Accurate flow cytometric membrane potential measurement in bacteria using Diethyloxacarbocyanine and ratiometric technique. Cytometry 1999; 35: 55–63.
23. Packard T. The measurement of respiratory electron-transport activity in marine phytoplankton. J Mar Res 1971; 29: 235–244.
24. Packard T, Denis M, Rodier M, Gardfield P. Deep-ocean metabolic CO_2 production calculations from ETS activity. Deep-Sea Res 1988; 35(3): 371–382.
25. Packard TT, Berdalet E, Blasco D, Roy SO, St-Amand L, Lagacé B, et al. CO_2 production predicted from isocitrate dehydrogenase activity and bisubstrate enzyme kinetics in the marine bacterium *Pseudomonas nautica*. Aquat Microb Ecol 1996a; 11: 11–19.
26. Packard TT, Berdalet E, Blasco D, Roy SO, St-Amand L, Lagacé B, et al. Oxygen consumption in the marine bacterium, *Pseudomonas nautica* predicted from ETS activity and bisubstrate enzyme kinetics. J Plankton Res 1996b; 18: 1819–1835.
27. Petit PX, Diolez P, De Kouchkovsky Y. Flows cytometric analysis of energy transducing organelles mitochondria and chloroplasts. In: Yen A (ed), Flow Cytometry: advanced research and clinical applications, pp 272–303. Florida: CRC Press, 1989.
28. Petit PX, O'Connor JE, Grunwald D, Brown SC. Analysis of the membrane potential of rat- and mouse-liver mitochondria by flow cytometry and possible applications. Eur J Biochem 1990; 194: 389–397.
29. Petit PX, Susin SA, Zamzami N, Mignotte B, Kroemer G. Mitochondria and programmed cell death: back to the future. FEBBS Lett 1996; 396: 7–13.
30. Ratinaud MH, Revidon S. A flow cytometric method to assess functionnal state of the *Listeria* membrane. J Microbiol Methods 1996; 25: 71–77.
31. Redfield AC, Ketchum BH, Richards FA. The influence of organisms on the composition of sea-water. In: Hill MN (ed), The Sea, pp 26–77. New York: Wiley-Interscience; 1963.
32. Rigoulet M, Guerin B, Denis M. Modification of flow-force relationships by external ATP in yeast mitochondria. Eur J Biochem 1987; 168: 275–279.
33. Ronot X, Benel L, Adolphe M, Mounolou JC. Mitochondrial analysis in living cells the use of rhodamine 123 and flow cytometry. Biology of the Cell 1986; 57: 1–8.
34. Rottenberg H, Wu SS. Quantitative assay by flow cytometry of the mitochondrial membrane potential in intact cells. Biochimica et Biophysica Acta 1998; 1404: 393–404.
35. Roy SO, Packard TT, Berdalet E, St-Amand L. Impact of acetate, pyruvate and physiological state on respiration and respiratory quotients in *Pseudomonas nautica*. Aquat Microb Ecol 1999; 17: 105–110.
36. Sabnis RW, Deligeorgiev TG, Jachak MN, Dalvi TS. $DiOC_6(3)$: a useful dye for staining the endoplasmic reticulum. Biotechnic & Histochemistry 1997; 72(5): 253–258.
37. Shapiro HM. Practical flow cytometry. New York: Liss AR, Inc; 1988.
38. Shapiro HM. Membrane potential estimation by flow cytometry. Methods 2000; 21: 271–279.
39. Takahashi T, Broecker WS, Langer S. Redfield ratio based on chemical data from isopycnal surfaces. J Geophys Res 1985; 90: 6907–6924.
40. Vives-Rego J, Lebaron P, Nebe-Von Caron G. Current and future applications of flow cytometry in aquatic microbiology. FEMS Microbiol Rev 2000; 24: 429–448.
41. Williams PJLB. Heterotrophic bacteria and the dynamics of dissolved organic material. In: Kirchman DL (ed), Microbial ecology of the oceans, pp 153–200. New York: Wiley-Liss, Inc.; 2000.

Author for correspondence: Michel Denis, Laboratoire d'Océanographie et de Biogéochimie, Université de la Méditerranée, CNRS UMR 6535, 163 avenue de Luminy, Case 901, 13288 Marseille cedex 9, France
Phone: +33 4 91 82 91 14; Fax: +33 4 91 82 65 48
E-mail: denis@com.univ-mrs.fr

Methods in Cell Science **24**: 107–114 (2002).

Cellular and serological markers of disease activity in Indian patients with HIV/AIDS

S. Sehgal, U. Datta, S. Mujtaba, A. Sood and V. K. Vinayak
Departments of Immunopathology & Internal Medicine PGIMER, Chandigarh and Department of Biotechnology, New Delhi, India

Abstract. There has been an exponential rise of HIV positive patients as observed at the surveillance center of Nehru Hospital. Most patients are poor and cannot afford repeated viral load assays. Therefore, there is a need to identify cost effective and reliable surrogate markers of disease activity. In the present study absolute number of CD4 cells, β_2 microglobulin, circulating necleosomes were studied in 30 patients of AIDS, 30 seropositives and 30 healthy controls. In addition viral load, P-24 assay, and TNFR-II assays were done in seropositive and AIDS patients.

The mean CD4 cells in patients with AIDS were 69.66 ± 68.25 mm^3 while in seropositives values was 370 ± 201.29 mm^3. The mean CD4 cells in healthy controls were however 690 ± 198 mm^3. The differences in all the groups were highly significant ($p < 0.001$). The mean CD4 values in Indians are significantly lower than reported from the west. The lower number of CD4 cells in healthy population is interpreted to be due to immune activation. The CD8 cell number in controls was 650 ± 207 mm^3 this figure is also higher than that observed in the west. P-24 assay failed to delineate between seropositives and patients with AIDS. Although, β_2 microglobulin levels were significantly higher in AIDS than in seropositives and higher in seropositives than in controls yet with the best possible cut off, it had a sensitivity of only 70% in delineating the two conditions. The correlation between CD4 cells and viral load was more significant when the CD4 cells were below 200 mm^3. Five out of 30 patients with a CD4 of 300–600 mm^3 had a viral load of over 1×10^5 cop/ml. The difference in TNF R-II levels between seropositives and AIDS was however more impressive. With a cut off of 550 pg/ml it had a sensitivity of 95% in delineating HIV from AIDS. It is concluded that a combination of absolute number of CD4 cells and TNF R-II assay along with clinical evaluation may be used to monitor therapy in resource poor countries where frequent viral load assay is unaffordable.

Key words: AIDS, CD4 cells, Surrogate markers, TNF, Viral load

Introduction

Acquired Immunodeficiency syndrome has proved to be the greatest challenge to biomedicine of 20th century [4, 6, 25]. According to UNAIDS, 42 million people are living with AIDS globally while 21 million have died of this deadly disease [31].

The National AIDS Control Organization Of India (NACO) estimates that 4 million people are infected In India [23]. The presence of anti HIV antibodies to various viral proteins is the key evidence of infection past or present [12]. Since the virus gets integrated into the host cell DNA, antibodies persist for lifetime. It is thus difficult at times to delineate active disease or AIDS from latent infection (seropositivity). Several surrogate markers have been investigated from time to time. T helper cells are the key targets of a suicidal infection by HIV virus [10]. A decrease in the absolute number of CD4 cells has been consistently shown to be associated with disease progression in all risk groups in all stages of HIV and in all ethic groups [7, 26, 30]. The basal levels however may differ in different ethnic groups [18].

There are however studies conducted over the past 5 years which indicate that CD4 counts may be normal in a few patients even in the presence of active viral replication and conversely CD4 counts may be very low in seropositive patients with tuberculosis and thus estimation of viral load is the gold standard for monitoring disease activity [16, 17, 27]. But in resource poor countries, both viral load assay as well as study of repeated CD4 estimations are unaffordable and thus there is a need to identify other reliable surrogate markers of disease activity and progression.

A part of the core protein i.e. P-24 is shed into the circulation during active viral replication and it correlates inversely with CD4 number [2, 20]. Further, it is also positive in the window period when the ELISA test is negative. The other surrogate markers that have been used by different investigators are neopterine levels, β_2 microglobulin levels and Tumor Necrosis Factor receptor II (TNFR II) assays

[3, 8, 11]. Since these are ELISA based assays, easy to perform and cost effective, they may be more promising. However, their utility may vary in different populations. In the present study CD4, P-24, β_2 microglobulin levels & TNFR II levels have been estimated in 30 HIV seropositive patients and 30 patients with full blown AIDS to assess their predictive value in delineating the two conditions. Viral load assay was done in patients with HIV and AIDS and correlated with CD4 cells and TNFR-II assay.

Methods

Cases reporting to the AIDS Surveillance Center, Department of Immunopathology PGIMER, were included in the study. The diagnosis of HIV infection was established by two different ELISA tests, i.e. Recombigen (Cambridge Biotech, Ireland); Detect (Biochem Immunosystem Inc, Canada) and a rapid test, Capillus (Cambridge Biotech, Ireland) or Immunocomb HIV Bi spot (Orgenics, Israel). If all the three test were frankly positive, cases were designated as positive. In case of even a minor discord, Western blot test was performed (Newlav Blot – I/II, Sanofi Diagnostic, Pasteur, France).

Clinical diagnosis of AIDS was established by CDC criteria and a CD4 cell count less than $200/mm^3$. Thirty healthy laboratory controls were also included in the study and subjected to similar investigations except that CD4 assay was done in 40 subjects.

1. Estimation of CD_4+ T- Lymphocytes:
   Absolute CD_4+T – lymphocyte count was measured in a Becton Dickinson Flowcytometer using anti-CD_4CD_8 antibodies and consort – 30 programmes initially. However, in the latter part of the study, cell quest programme was used. Briefly, 100 μl of EDTA blood was incubated with 20 μl of the simulset reagent for 30 minutes. The mixture was subjected to red cell lysis using a 1 to 10 dilution of the lysing fluid (Becton Dickinson). After incubating for 10 minutes, the cells were washed thrice in phosphate buffer saline pH 7.2 and viewed in the flowcytometer. The percentage of CD_4+T-lymphocytes was enumerated and their absolute number was calculated from absolute number of total lymphocytes. In 10 cases of AIDS activation markers were also assessed by enumerating cells co expressing CD8/CD57 and CD8/CD38.
2. β_2M was estimated using an enzyme immunoassay (United Biotech, Inc, USA) based on a sandwich Enzyme – Linked Immunoosorbent Assay (ELISA) principal. Briefly, β_2M kit employs micro well coated with anti-β_2M antibodies. Patients sera diluted 1:1000 and standards were incubated followed by incubation with enzyme conjugated second antibody. The color was developed using tetramethyl benzidine hydrochloride (TMB). The intensity of the color was read at 450 nm. β_2M was quantitated in μg/ml by plotting standards Vs. O.D. on a double log graph paper according to the manufacturer's instructions.
3. P-24 assay was conducted using an Ortho-diagnostic kit according to manufacturers instructions. The sera were acid dissociated first and then subjected to an indirect ELISA. The P-24 levels in patients were calculated from the standard curve and expressed in pg/ml.
4. Tumor Necrosis Factor Receptor II Assay was based on a sandwich ELISA System (Immunotech, France). Soluble TNFR II was captured on the wells of a microtiter plate by a monoclonal antibody. The unbound TNFR II was washed out and a second biotinylated antibody was allowed to interact with antigen – antibody complex. In the third step avidin labeled peroxidase was added and incubation carried out. The color was developed with TMB and reading taken at 450 nm. The value was deduced from a standard curve and expressed as pg/ml.
5. Viral Load Assay
   HIV-RNA was estimated in the plasma of AIDS patients using Amplicor HIV-I monitor test (Roche Diagnostics, NJ, USA), according to manufacturers instructions. RNase free pipettes and tips were used through out the procedure. The test is based on extraction of RNA from plasma, reverse transcription – polymerase chain reaction (RT-PCR), followed by quantitative ELISA. Briefly, RNA was extracted using lysing reagent containing manganese chloride (Mncl) and quantitation standard (QS). The extracted RNA was precipitated in 100% isopropanol and then in 70% ethanol. The precipitate was dissolved in sample diluent. Fifty microlitre of patient RNA and control were taken for RT-PCR and mixed with 50 μl of master mix containing manganese solution. PCR was set up using following protocol.
   - Hold 2 min at 50 °C
   - Hold 30 min at 60 °C
   - A cycle – 10 Sec 95 °C, 10 Sec 55 °C and 10 Sec 72 °C × 4 cycles
   - B cycle – 10 Sec 90 °C, 10 Sec 60 °C and 10 Sec 72 °C × 26 cycles
   - Hold 15 min at 72 °C

   The amplified denatured products including standards were hybridized with biotin labeled probe coated on the micro wells of the ELISA plate and further incubated with avidin conjugated HRP and color developed with TMB. The viral load was calculated according to manufacturers formula.
6. Apoptosis
   Levels of circulating nucleosomes were estimated

in the sera of patients and controls using an ELISA kit (Boehringar Mannheim) the mean O.D. values of controls were compared with those of the patients. In 10 patients showing a large number of double negative cells, annexine V staining was also done and the intensity of fluorescence measured in a Becton Dickinson FACS scan.

Results

There has been an exponential rise in overall HIV positive cases as detected at the Surveillance Centre over the last 15 years in spite of aggressive information Education and Communication (IEC) program launched by NACO [16]. During 1987 to 1996, 622 seropositives were documented with 210 having full blown AIDS while in 2001 alone, 439 patients were documented (Figure 1). Thirty patients each of AIDS and HIV seropositives were randomly selected and studied in detail. There were 44 males and 16 females amongst the AIDS patients, tuberculosis was the single most important bacterial infection. Pulmonary tuberculosis with empyema in 2, pulmonary plus CNS TB in 1, disseminated TB in 7 and lymph node TB in 3. Three patients each had cryptococcal meningitis and repeated pneumonia. Herpes Zoster was noted in two cases. Others had Nocardia, Histoplasma, secondary syphilis, molluscum contagiosum etc.

Absolute CD4 cell number

The mean value in the AIDS group was 69.66 ± 68.25 mm^3 with values ranging from < 10 mm^3 to 154 mm^3. The absolute CD4 cell number in seropositive on the other hand was 370 ± 201.29 mm^3 with values ranging from 239 to 589 mm^3. The results were highly significant ($p < 0.001$) the mean CD4 value in the healthy control group was 690 ± 198 mm^3 with a range of 232 to 1,075 mm^3. The difference between the control group and seropositive were also highly significant ($p < 0.001$). It is important to note that the mean CD4 cell number in healthy control was lower than reported from the west (Figures 2–4 and Table 1).

β_2M assay

Out of the other more economical surrogate markers, β_2 microglobulin was studied in all the patients and controls. The levels of β_2 microglobulin were statistically higher in patients with full blown AIDS than in seropositives ($p < 0.05$). The mean value in patients with AIDS was 6.7 ± 2.3 μg/ml with a range of 11 μg/ml to 1 μg/ml. The β_2M values in the seropositives ranged between 9 μg to < 1 μg with a mean of 4.15 ± 2.17 μg/ml. Mean β_2M levels in healthy controls were 1.7 ± 3.5 μg/ml .The difference between the control group and seropositives was also significant ($p < 0.05$). If one considers a level of 5.2 μg as the cut off, it had a sensitivity of only 70%. Since there was a significant overlap between the values in the two groups, β_2M appears to be a poor predictor of disease activity in a given patient.

P-24 antigenemia

P-24 antigenemia assay was conducted using a new Orthodiagnostic kit to establish its significance in delineating seropositive cases from those with full-

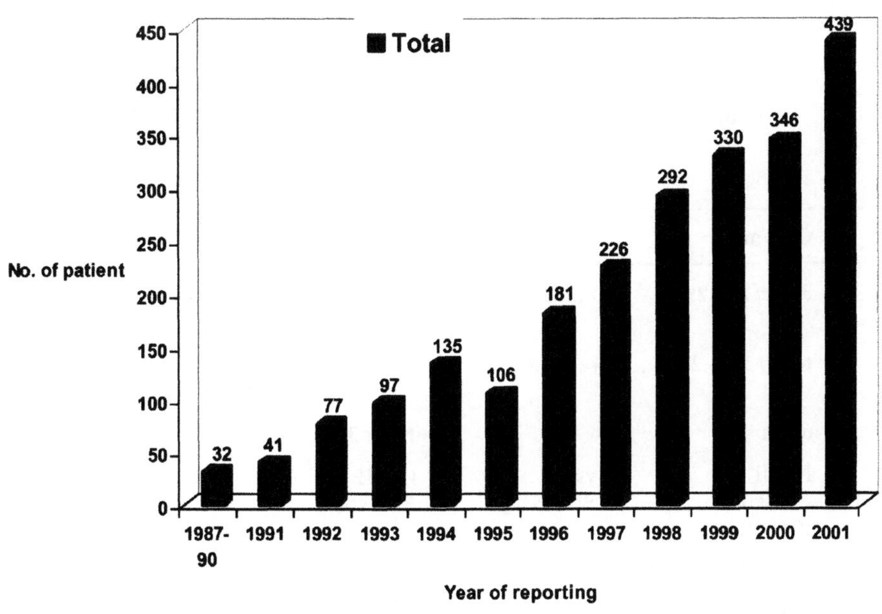

Figure 1. Time Trends of HIV Infection as observed at Surveillance Center, Chandigarh.

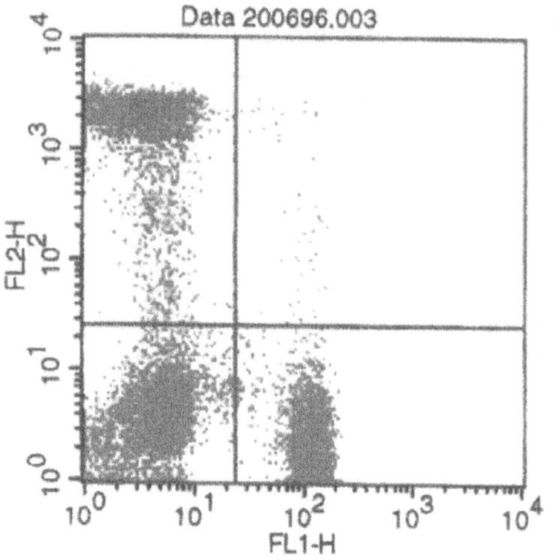

Figure 2. CD4 & CD8 cells in healthy laboratory subjects showing increased CD8 cells with a CD4: CD8 ratio of 1.1.

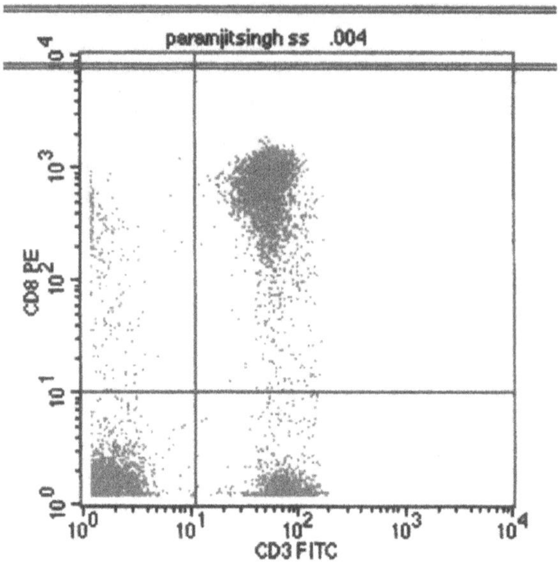

Figure 3. Marked increased in CD8 cells in an HIV positive patients.

A

B

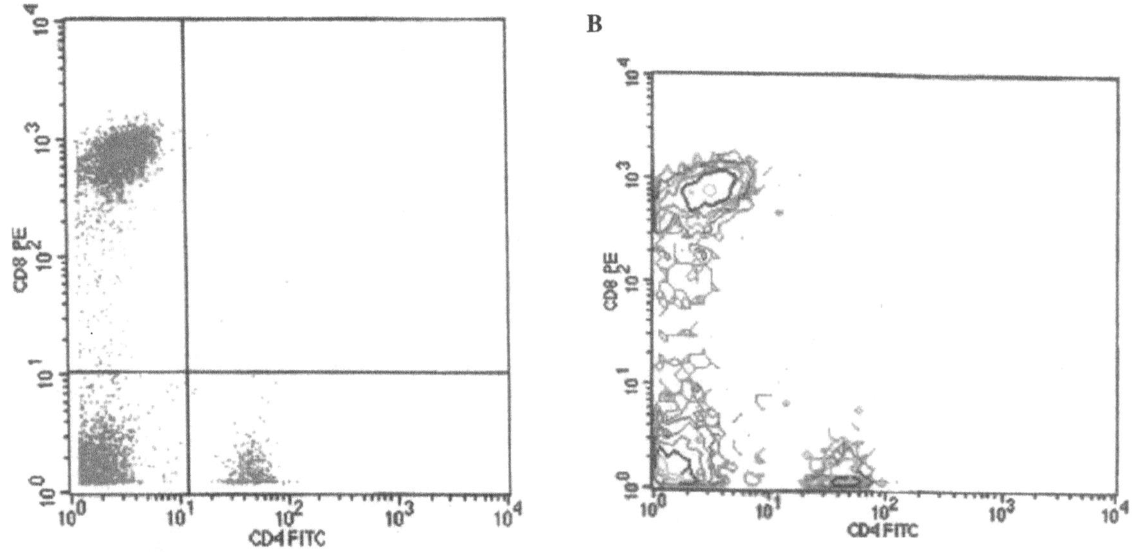

Figure 4. Patient with AIDS showing gross lowering of CD4 cells.

Table 1.

Groups	CD4/mm^3	P-24 pg/ml	β_2M μg/ml	TNFR-II pg/ml	Viral load
Seropositive I	370 ± 201.29	40 pg ± 70.30	4.15 ± 2.17	280 ± 154	0.56.23 × 10^5 ± 81 × 10^5
AIDS II	69.66 ± 68.25	44 ± 80.42	6.7 ± 2.3	799 ± 285	6.63 ± 10^5 5.78 × 10^5
Control III C	690 ± 198	–	1.7 ± 3.5	–	–
Signifance	I vs II $P < 0.001$ II vs III $P < 0.001$ I vs III < 0.001	1 vs 2 $P > 0.05$	I vs II $P < 0.05$ I vs III $P < 0.05$ II vs III $P < 0.05$	I vs II $P < 0.001$	I vs II $P < 0.001$

blown AIDS. The mean P-24 value in full-blown AIDS was 44 pgm/ml with a range of 4–200 pg/ml while in seropositives it was 40 pg/ml with a range of 4–200 pg/ml the difference in the two groups was insignificant ($p > 0.05$) There was also a poor correlation between CD4 number and P-24 levels ($r = 0.23$).

TNF receptor R II assay

It is well documented that TNF levels are elevated in AIDS, which is partly responsible for the wasting. Further, soluble TNFR-II has been adjudged as a suitable surrogate marker of disease activity. Studies conducted at our center revealed mean TNF R II levels of 799.13 ± 285.18 pg/ml in patients with full blown AIDS while in seropositives the mean value was 280 ± 154.92 pg/ml only. The difference was highly significant ($p < 0.001$) Thus, TNF R-II assay appears to be a reliable cost effective surrogate marker of disease activity. Soluble TNF R-II levels were inversely correlated with CD4 cell ($r = 0.71$).

Viral load

The viral load in patients ranged from 22×10^5 to 80,000 copies/ml with a mean of $6.63 \times 10^5 \pm 5.78$ copies/ml. The viral load in seropositives ranged from 2.3×10^5 to 370 copies/ml with a mean of $0.56 \times 10^5 \pm 0.81 \times 10^5$ copies/ml. In 10 percent of the patients the viral load was more than 1×10^5 although the patients were asympomatic and had a CD4 counts of > 300/mm^3. There was however a clear cut inverse correlation between the CD4 cells and the viral load particularly when the CD4 cells were below 200 (Figure 5).

Circulating nucleosomes

The mean OD value of nucleosomes in AIDS was 2.99 ± 0.32 while in seropositive group it was 1.22 ± 0.63. In the healthy controls the ELISA value was 0.256 ± 0.09. The difference between the AIDS and seropositive groups was highly significant ($p < 0.01$). Using FACS analysis and staining with annexure V, eight out of ten patients studied (with a CD 4 count of less than 100) gave a strong staining with annexure V with more than 30% cells undergoing apoptosis (Figure 6).

Discussion

The various surrogate markers of disease activity that have been used from time to time include absolute number of CD4 cells, β_2 microglobulin, neopterine Levels, P-24 antigen assay and TNF R II Assay.

CD4 and CD8 cell numbers

Historically, absolute number of CD4 (helper) T cells has been used as a reliable indicator of disease progression. These are the key targets of a suicidal infection by HIV virus. In one study comprising a population of uninfected homosexual men, the absolute number, the percentage and CD4: CD8 ratios were 898 ± 318 mm^3, 44.5 ± 8.2% and 1.58 ±

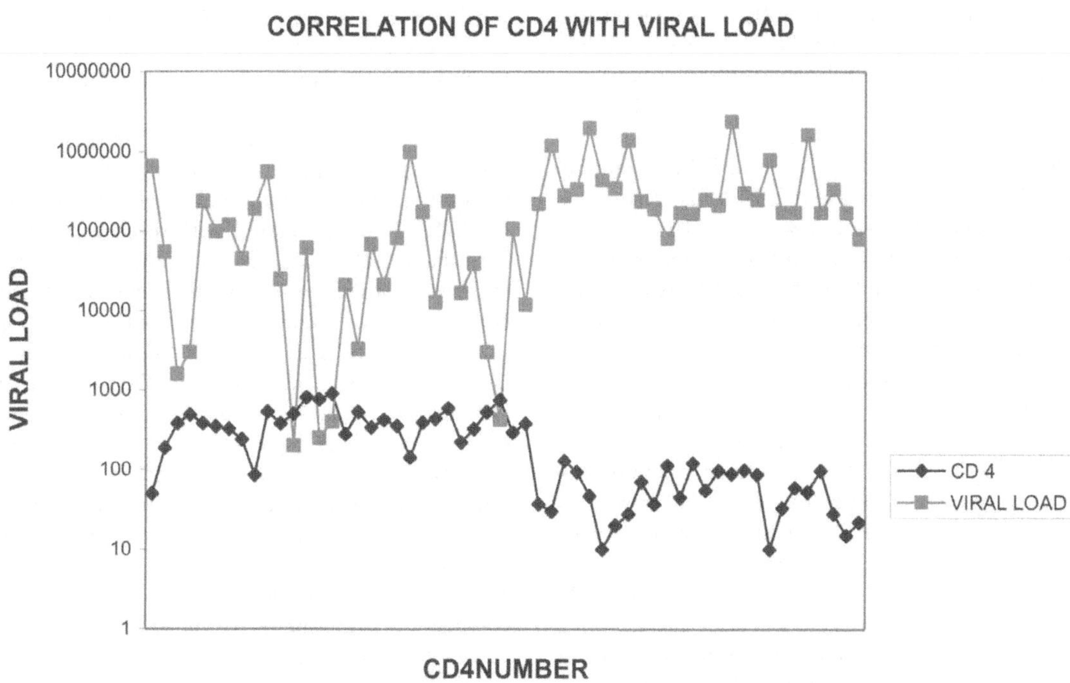

Figure 5. Increase correlation between CD4 cells and viral load in patients with HIV/AIDS.

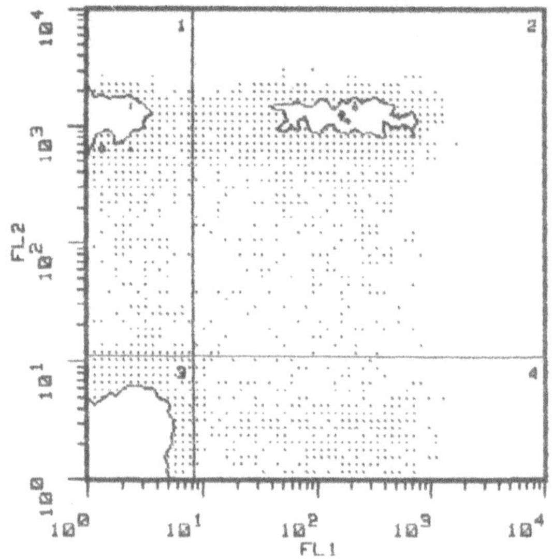

Figure 6. Dual expression of CD8 ± 38 in-patient with AIDS.

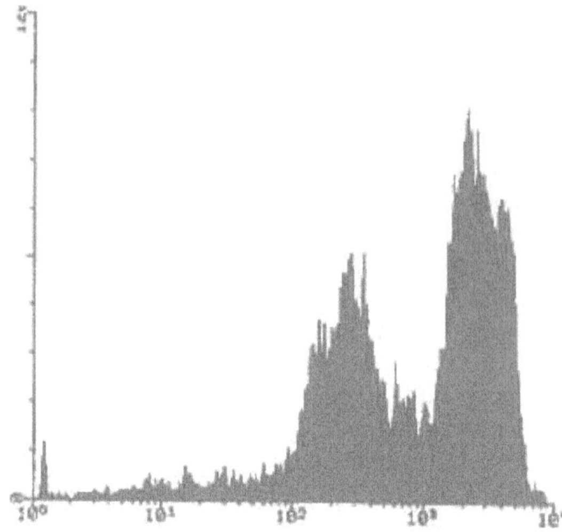

Figure 7. Annexin V staining in a patient with AIDS apoptotic cells.

0.66 respectively [30]. CD4 T cells levels below 200 mm^3 were documented to be associated with AIDS defining opportunistic infections and have been used for prophylaxis. The correlation was so impressive that center for disease control identified levels below 200 mm^3 as indicator of full- blown AIDS [7, 19]. The levels generally fall below 50 preterminally.

Absolute number of CD4 helper cells have been consistently shown to be associated with disease progression Among homosexual men [13a, 13b] those with less than 200/mm^3 CD4 cells had a 70% chance of developing AIDS within 3 years as opposed to 11% in those with CD4 count of more than 500 mm^3. Taylor et al. [30] documented that CD4 cells percentage and CD8 ratios were slightly more prognostic than CD4 cell number alone.

In the present study the mean CD4 cell in 40 healthy controls were 690 ± 198 mm^3 while in patients with full blown AIDS the CD4 cell number was 69.66 ± 68.25 mm^3 and in healthy HIV seropositive individuals the values were 370.01 ± 201.29 mm^3. Kelinkovich [18] and colleagues reported that helminthes and other infections could be responsible for lower T cell number in their controls because of sustained immune activation.

Giorgi et al. [13] documented that decline in CD4 cells occurs in 4 stages. Over the first 12–18 months there is a rapid drop. In the classical Multicenter AIDS Study (MACS) the CD4 number of 945 during the last seronegative stage dropped to 740 mm^3, 697 mm^3 and 647 mm^3 at 6, 12, 18 months respectively.

A transient though more profound drop occurs during seroconversion but the levels bounce back. Stage II is characterized by a stable of slow decline where levels stabilize to an average of 500 mm^3. In long-term non-progressors, it may last 15–20 years. In Stage III, which begins about 2 years prior to clinic manifestation of AIDS, is characterized by a sharp decline. Stage 4, observed during terminal period is characterized by a slower decline of CD4 cells; a small percentage of patients can live for months or years with no detectable levels of CD4.

When CD4 cells are below 200 mm^3, correlation with disease progression is more impressive as documented by Fahey et al. [8]. Our observations on viral load and TNF assays corroborate these findings.

The number of CD8 cells in healthy Caucasians is 500 mm^3. At the time of infection the levels increase presumably as an indicator of immune activation and the levels may rise as high as 1,000 mm^3. About 2 years prior to AIDS development the levels drop to basal levels (22). In our study, the mean CD 8 cells in controls were higher than those reported from the west. (mean 650 ± 207.03). Further, certain immune activation markers have been linked to disease activity. It has also been shown by Giorgi et al. [14, 15] that dual expressions of CD8/38 or CD8/57 are associated with disease progression. We have also confirmed above observation (Figure 7) however due to financial and technical constraints they cannot be used routinely. A population of CD8 cells that express high levels of HLA DR and CD38 is observed in patients with poor prognosis. It has been suggested that precocious killer CD8 cells can lyse infected or uninfected CD4 cells, secondly high CD38 expression may represent immunological immaturity of CD8 cells and finally CD8/38 positive cells may reflect a cytokine driven response to high viral load [24]. The CD4 cell are also more susceptible to apoptosis (24) further leading to lymphopenia Eight patients in terminal stages showed a large number of unstained cells; when these were subjected to annexin V staining revealed intense staining in more than 30% of cells.

Similarly levels of nucleosomes were significantly higher in AIDS.

Beta 2 microglobulin and P-24 assays

Beta 2 microglobulin is a non-convalently linked light chain component of HLA class I molecules present on most nucleated cells and lymphocytes express high concentration of β_2M on the surface and release it during immune activation via cytokines stimulation or cell destruction.

Circulating levels of β_2M are thus controlled by turnover of nucleated cells renal function and immune activation. Raised levels of β_2M have been documented in HIV/AIDS infections etc. In multiple myeloma, it is used as a reliable prognostic marker of disease activity [5]. While β_2M was significantly higher in AIDS than in seropositives ($p < 0.05$) and higher in seropositives than in control ($p < 0.05$), yet when 5.2 μg/ml was considered as the cut off, it had a sensitivity of 70% in delineating HIV from AIDS. Thus, in a given case it has a limited value.

P-24 assay used in the study also failed to delineate HIV from AIDS in our population although Allian et al [1] documented that P-24 antigenemia is a better predictive marker of HIV related infections, thrombocytopenia and severe disease. Matheisen et al. [20] advised acid dissociation of samples. In our study, even after acid dissociation we could not obtain statistically significant results. It appears that P-24 complexed with antibody could not be readily dissociated even when acid dissociation is employed. Thus it is not a useful marker of disease activity in Indian patients. Similar observations have earlier been made in black population in the west in hemophiliacs.

Neopterine levels

Neopterine is one of the family of pteridines that are derived in vivo from guanosine triphosphate (GTP) and is known to reflect immune cells activation. Neopterine production rises in lymphocytes and macrophage in inflammatory conditions of viral, bacterial, protozoal origin, graft versus host reaction and graft rejection

It has been extensively used as a surrogate marker of disease activity in HIV/AIDS in the west. Melmed et al. [21] reported a strong inverse correlation between rise in neopterine levels and fall in CD4 cells number in the first year following HIV infection in a cohort of homosexual males. We did not estimate neopterine levels. It must be stressed that high neopterine levels have been documented in various infections and since in India, at least 50% of patients with AIDS have tuberculosis as a co-infection it may lose its fidelity as a marker of disease activity.

Tumor Necrosis factors (TNFα)

Tumor necrosis factor is produced by a number of cells notably by the monocyte macrophage system. The levels are reported to be significantly higher in AIDS than in controls or seropositives and therefore TNFα receptor II (TNF α II) has been used as a marker of disease activity. In this study, the mean TNFα R II levels in AIDS were 799.13 pg/ml while in seropositives the values were 285.18 pg/ml. The difference was highly significant ($p < 0.001$). There was also a significant positive correlation with viral load. Thus a combination if CD4 assay and TNF R II assay could be used as reliable markers of disease activity. Infact raised levels are so impressive as to warrant novel therapeutic strategies to inhibit TNF production in AIDS especially in 'slim disease'. Interestingly thalidomide is a potent inhibitor of TNFα [28] and currently therapeutic trials are on in AIDS patients with wasting.

Acknowledgements

The study was funded by the Department of Biotechnology, Government of India, New Delhi. Authors feel grateful to Mr. I. James and N. Pasricha.

References

1. Allain JP, Laurian Y, Paul DA, Senn D, Members of the AIDS hemophilia French Study Group (1986). Serological markers in early stages of human immunodeficiency virus infection in haemophiliacs. Lancet ii: 1233–1236.
2. Allain JP, Laurian Y, Paul DA, Berroust F, Leuther M, Gazengel C, Senn D, Larrieu MJ, Bosser C (1987). Long-term evaluation of HIV antigen and antibodies to p24 and gp41 in patients with hemophilia. N Engl J Med 317: 1114–1121.
3. Anderson RE, Lang W, Shikoski S, Royce R, Jewelb N, Winkelstein W (1990). Use of β_2M and CD4 count to predict development of AIDS in persons with HIV infection. Arch Intern Med 150: 73–77.
4. Barre-Sinoussi F, Chermann JC, Rey F. et al. (1983). Isolation of a T lymphotrophic retrovirus from a patient at risk for acquired immune deficiency syndrome (AIDS). Science 220: 868–871.
5. Bataille R, Grenier JJ, Sany J (19XX). β_2M in myeloma: optimal use for staging, prognosis and treatement – A prospective study of 160 patients. Blood 63(2): 468–476.
6. Center for Disease Control (1981). Pnemocystis pneumonia. Los Angles Morbid-Mortal weekly Rep 30: 250–253.
7. Centers for Disease Control and Prevention (1992, 1993). Revised classification system for HIV infection and expanded surveillance case definition for AIDS among adolescents and adults, MMWR 41 (No. RR-17): 1–19.

8. Fahey JL, Taylor JMG, Detels R, et al. (1990). The prognostic value of cellular and serologic markers in infection with human immunodeficiency virus type-1. N Engl J Med 322: 166–172.
9. Fahey JL, Taylor JMG, Manna B, Nishanian P, Aziz B, Giorgi JV, Detels R (1998). Prognostic significance of plasma markers of immune activation, HIV viral load and CD4 T-cell measurements. AIDS 12: 1581–1590.
10. Fauci AS, Lane HC (1998). Human immunodeficiency virus disease; AIDS and related disorders. In: Fauci AS, Braunwald E, Isselbacher KJ, Wilson JD, Martini JB, Kasper DL, et al. (eds), Harrison's Principles of Internal medicine, 14th edn, p. 1829. New York, McGraw-Hill.
11. Fuchs D, Kramer A, Reibnegger, et al. (1991). Neopterin and β_2M as prognostic indices in HIV infection. Infection 19 (Suppl 2): S98–S102.
12. Gallo D, Diggs JL, Shell GR, et al. (1986). Comparison of detection of antibody to the acquired immune deficiency syndrome virus by enzyme immunoassay, immunofluorescence, and Western blot methods. J Clin Microbiol 23: 1049–1051.
13A. Girogi JV, Detels R (1989). T cell subsets in HIV infected homosexual men NIAID Multi center AIDS Cohort Study. Clin Immun Immunopathol 52: 10–18.
13B. Giorgi JV (1992). CD4 counts in relation to markers of immue activation. In: Janossay G, Autran B, Miedema F (eds), Immunodeficiency in HIV infection and AIDS, pp 1–17. Karger, Basel.
14. Giorgi JV, Liu Z, Hultin LE, Cumberland WG, Hennessey K, Detels R (1993). Elevated levels of CD38+CD8+T cells in HIV infection add to the prognostic value of low CD4+T cell levels: Results of 6 years of follow-up. J AIDS 6: 904–912.
15. Giorgi JV, Ho HN, Hirji K, Chou CC, Hultin LE, O'Rourke S, Park L, Margolick JB, Ferbas J, Phair JP, and the Multicenter AIDS Cohort Study (1994). CD8+lymphocyte activation at HIV-I seroconversion: Development of HLA-DR+CD38-CD8+ cells is associated with subsequent stable CD4+cell levels. J Infect Dis 170: 775–781.
16. Ho DD, Mondgil T, Alan M (1989). Quantitation of human immunodeficiency virus type 1 in the blood of infected persons. N Eng J Med 321: 1621–1625.
17. Harrigan R (1995). Measuring viral load in the clinical setting. J Acqui Immune Defic Syndr 10: S34–S40.
18. Kalinkovich A, Weisman Z, Greenberg Z, Nahmias J, Eitan S, Stein M, Bentwich Z (1998). Decreased CD4 and increased CD8 counts with T cell activation is associated with chronic helminth infection. Clin Exp Immunol 114: 414–421.
19. Lang W, Perkins H, Anderson RE, Royce R, Jewell N, Winkelstein W (1989). Pattern of T lymphocyte changes with human immunodeficiency virus infection: From seroconversion to the development of AIDS. Journal of Acquired Immune Deficiency Syndromes 2: 63–69.
20. Mathiesen T, Sundqvist VA, Albert J, Ohlsson E, Wahren B (1988). Acid hydrolysis of serum samples to increase detection of HIV antigen. Journal of Virological Methods 22: 143–148.
21. Melmed RN, Tayor JMG, Detels R, Bozorgmehri M, Fahey JL (1989). Serum neopterin changes in HIV-infected subjects: indicator of significant pathology, CD4 T cell changes, and the development of AIDS. J.Acquired Immune Defic Syndr 2: 70–76.
22. Margolick JB, Munoz A, Donnenberg AD, Park LP, Galai N, Giorgi JV, O'Gorman MRG, Ferbas J, the Multicenter AIDS Cohort Study (1995). Failure of T-cell homeostasis proceeding AIDS in HIV-1 infection. Nature Med 1: 674–680.
23. National AIDS Control Organization India. Status and trend of AIDS epidemic India up to 2001 (http://naco.nic.in/vsnaco/indianscene/trend.htm).
24. Oyaizu N, Pahwa S (1996). Phenotype and function of T cells in HIV disease. Immunology of HIV infection Ed. Sudhir Gupta Planum Medical Boof Co N.Y. London, chapter 9, pp 133–159.
25. Popovic M, Sarangdharn MG, Read E, Gallo RC (1984). Detection, isolation and continuous production of cytopathic retroviruses (HTLV-III) from patients with AIDS. Science 224: 500–504.
26. Philips AN, Lee CA, Elford J, Janossy G, Timms A, Bofill M, Kernoff PBA (1991). Serial CD4 lymphocyte counts and development of AIDS. Lancet 337: 389–392.
27. Phillips AN, Eron JJ, Barlett JA, et al. (1996). HIV-I RNA levels and the development of clinical disease. AIDS 10: 859–865.
28. Rose NR, Leskovsek N (1998). Thalidomide in scleroderma: immunopathogenesis and treatment. Imm Today: 499–501.
29. Sud A, Sehgal S, Sharma M, Varma S, Kumar B, Sharma BK (1999). Spectrum of secondary infections associated with HIV infection in North India. Ann Natl Acad Med Sc India 35(3&4): 151–158.
30. Taylor JMG, Fahey JL, Detels R, Giorgi JV (1989). CD4 percentage, CD4 number, and CD4:CD8 ratio in HIV infection: which to choose and how to use. J AIDS 2: 114–124.
31. UNAIDS. www.unaids.org/epidemic_update/report_dec012.

Address for correspondence: Dr. Shobha Sehgal, Department of Immunopathology, PGIMER, Chandigarh, India-160012
Phone: 91-172744180 or 91-172-747585 (ext. 5194)
Fax: 91-172-745094; E-mail: shobhasehgal@yahoo.com

GPSR Compliance

The European Union's (EU) General Product Safety Regulation (GPSR) is a set of rules that requires consumer products to be safe and our obligations to ensure this.

If you have any concerns about our products, you can contact us on ProductSafety@springernature.com

In case Publisher is established outside the EU, the EU authorized representative is:

Springer Nature Customer Service Center GmbH
Europaplatz 3
69115 Heidelberg, Germany

Batch number: 10585315

Printed by Printforce, the Netherlands